JUMBO COLORING BOOK

COMMON NAME: WASPS

Hymenoptera is one of the largest and most well-known orders of insects, and it includes Wasps.

Biographical Note
The Nature of Wasps An Educational Coloring Book is a new work,
first published by Little Artist Studio in 2025.

International Standard Book Number
ISBN 979-8-9992504-6-9

www.littleartiststudio.org

Discover the fascinating world of wasps with over 90 detailed coloring pages. These incredible insects play vital roles in ecosystems, contributing to nature in many important ways. Part of Little Artist Studio's acclaimed educational series, each illustration tells a captivating story that inspires curiosity and creativity. Featuring single-sided pages, artists of all ages can use any coloring medium and easily showcase their finished masterpieces. Perfect for nature enthusiasts, educators, and creative minds alike.

SCIENTIFIC NAME FOR WASPS: HYMENOPTERA

Wasps are part of a big insect group called Hymenoptera, which also includes bees and ants. There are many kinds of wasps, and each kind has its own special scientific name.

LEARNING ABOUT WASPS

Entomology is the study of insects, and wasps are one type of insect. A wasp's life cycle has four main stages.

STAGE 1: EGG

The life cycle begins when a female wasp lays an egg, usually in a nest or inside a host (for parasitic wasps).

STAGE 1: EGG

Wasp eggs are small, white, and cylindrical

STAGE 1: EGG

The queen wasp lays a single egg in each cell of her nest, which hatches into a larva in 5-8 days

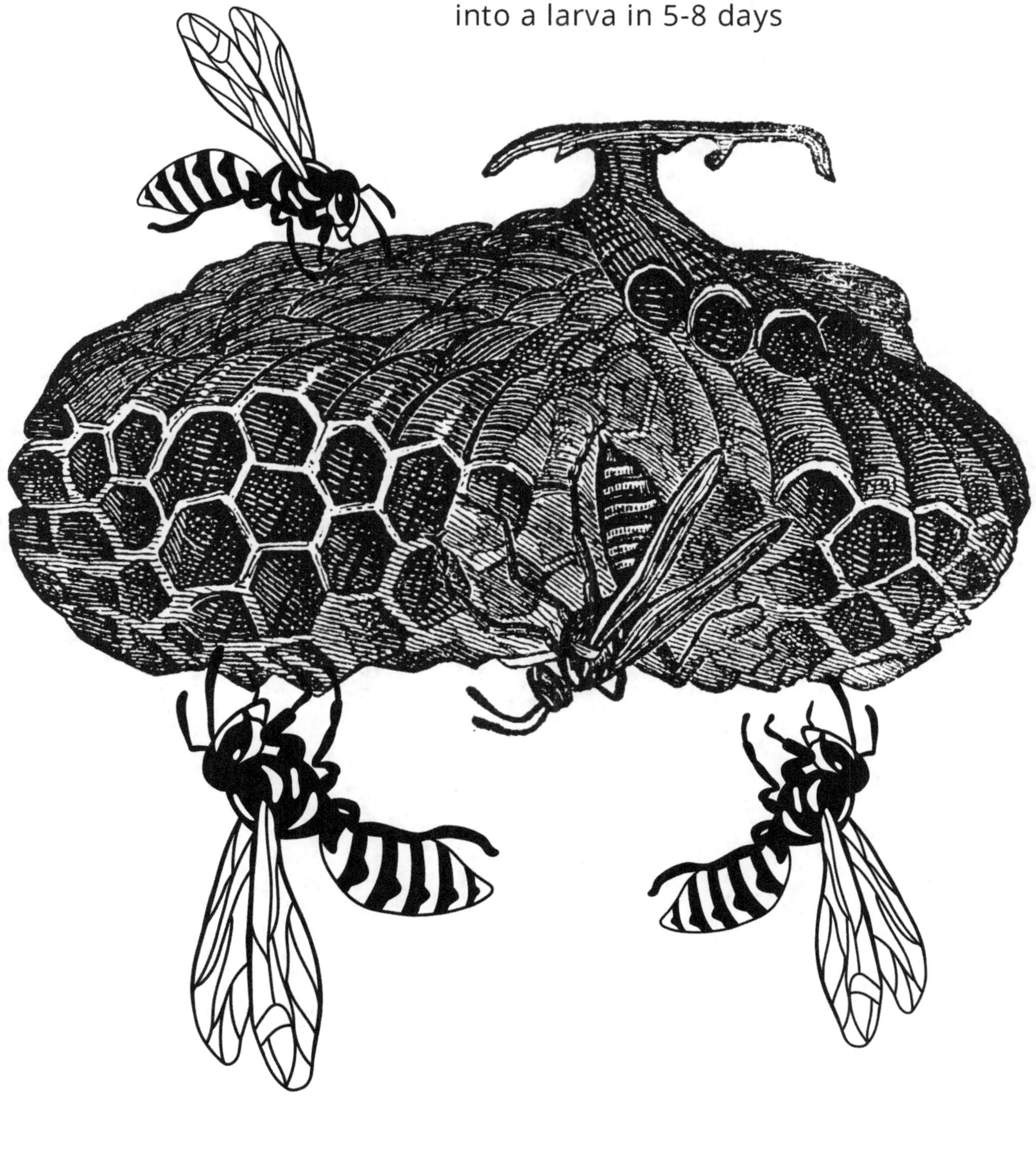

STAGE 2: LARVA

The egg hatches into a larva, which looks like a small, soft, white grub.

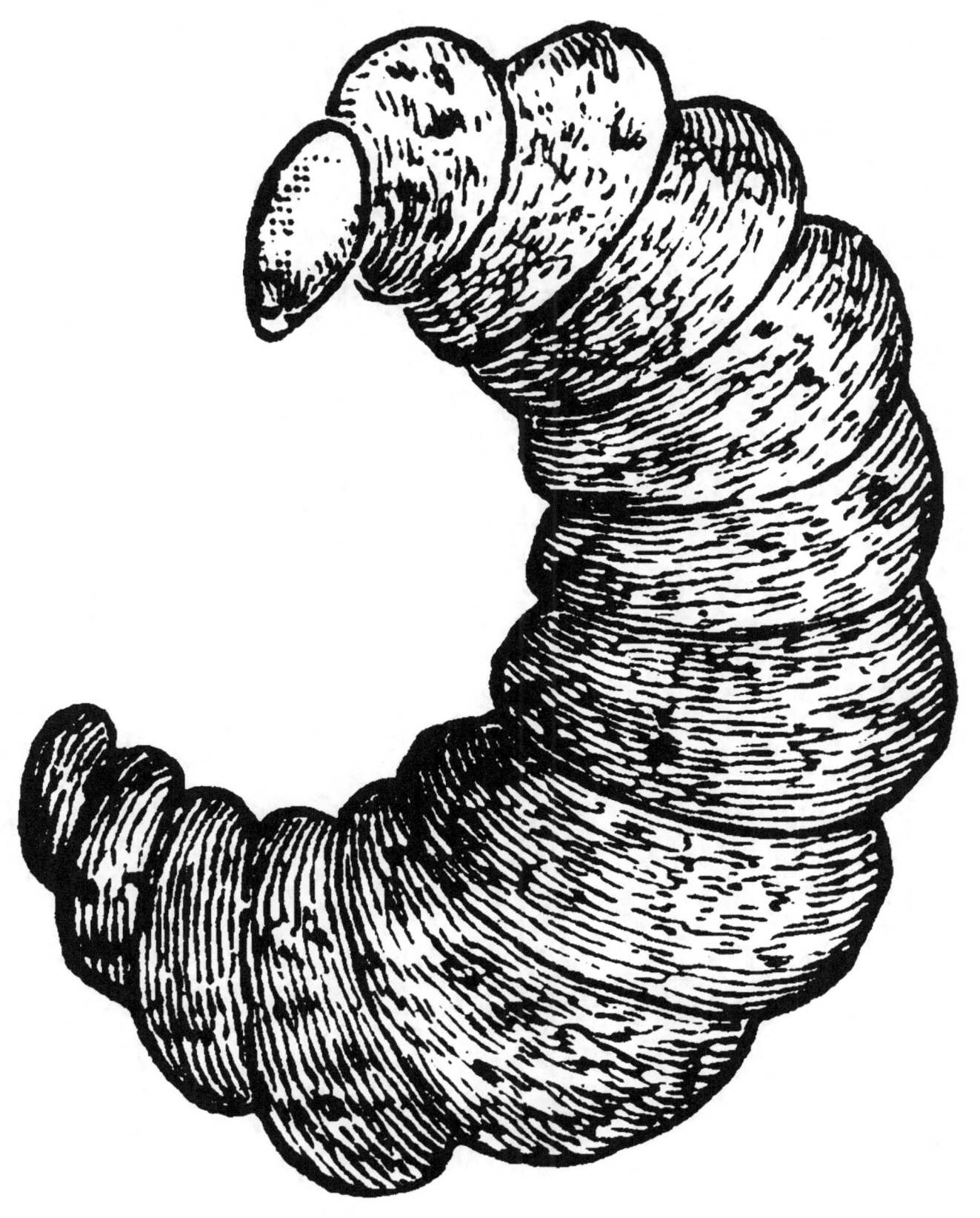

STAGE 3: PUPA

After growing enough, the larva spins a cocoon or forms a protective case and becomes a pupa.

STAGE 3: PUPA

Inside, it transforms into its adult wasp.

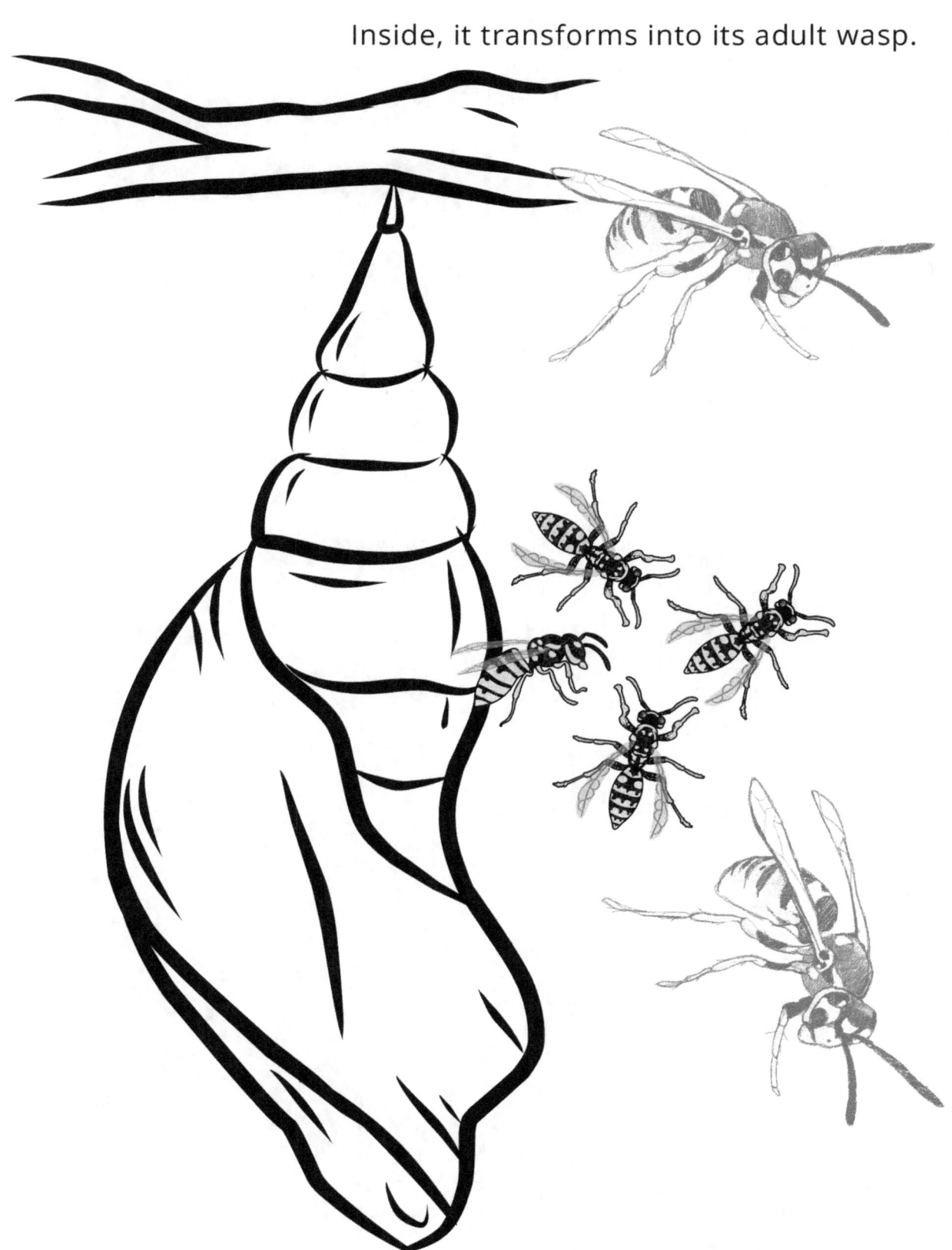

STAGE 4: ADULT

The adult wasp emerges from the pupa. It can now fly, find food, reproduce, and repeat the cycle.

WINGS

A wasp has two pairs of transparent wings.

QUEEN WASP

The queen wasp is the only one to survive the winter and starts a new colony in spring.

DIET: OMNIVORE

Wasps are omnivores, meaning they eat both plant and animal matter.

DIET: OMNIVORE

Adult wasps mainly eat sugars from nectar and fruit juices, while larvae feed on insects and other small animals.

DIET: OMNIVORE

Wasps hunt some insects and bugs, including flies and caterpillars to feed their larvae.

DIET: OMNIVORE

Wasps will also consume juices from overripe fruits.

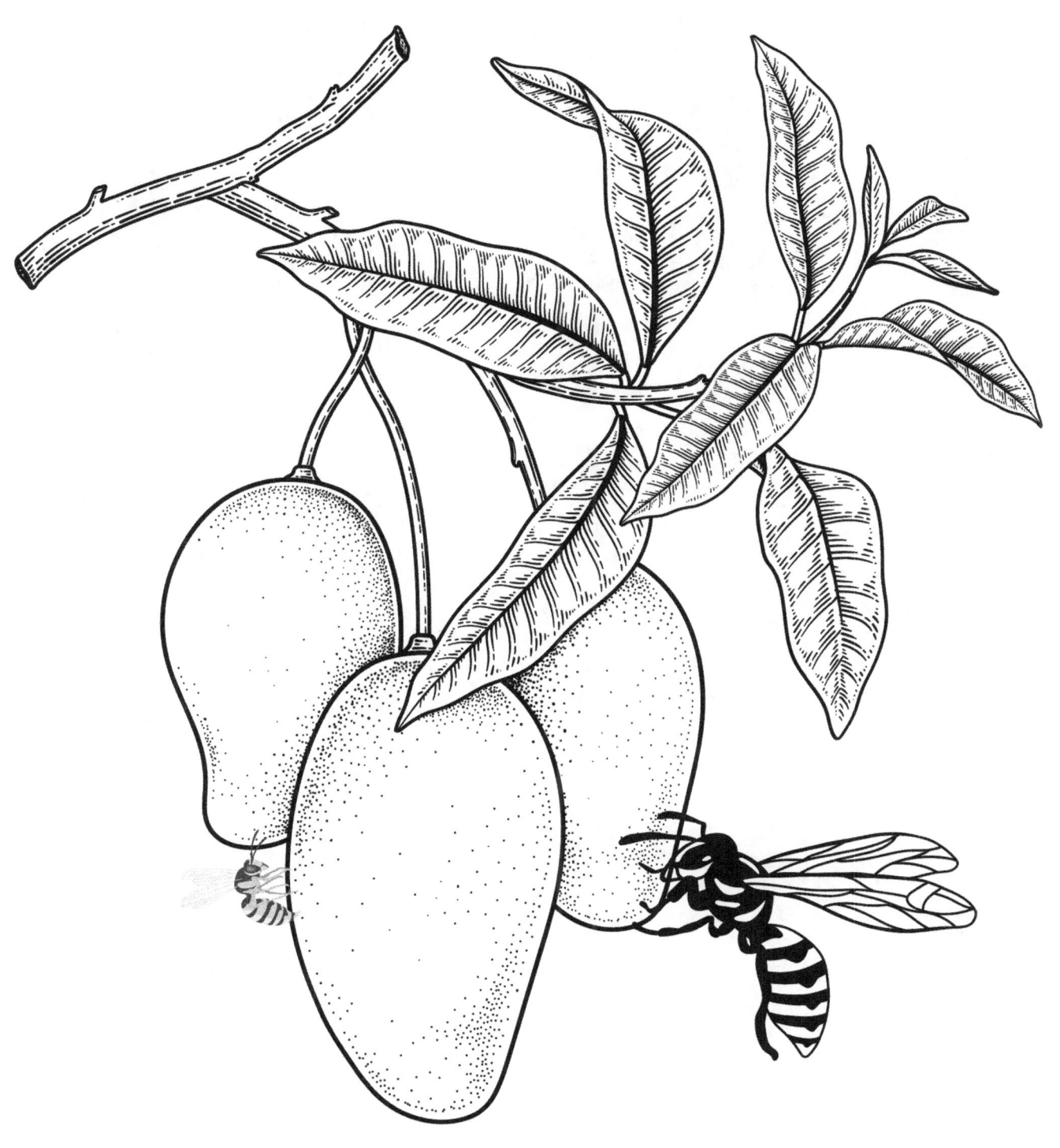

WASP SPECIES

There are over 30,000 described species of wasps, but scientists estimate that the actual number could exceed 100,000species worldwide.

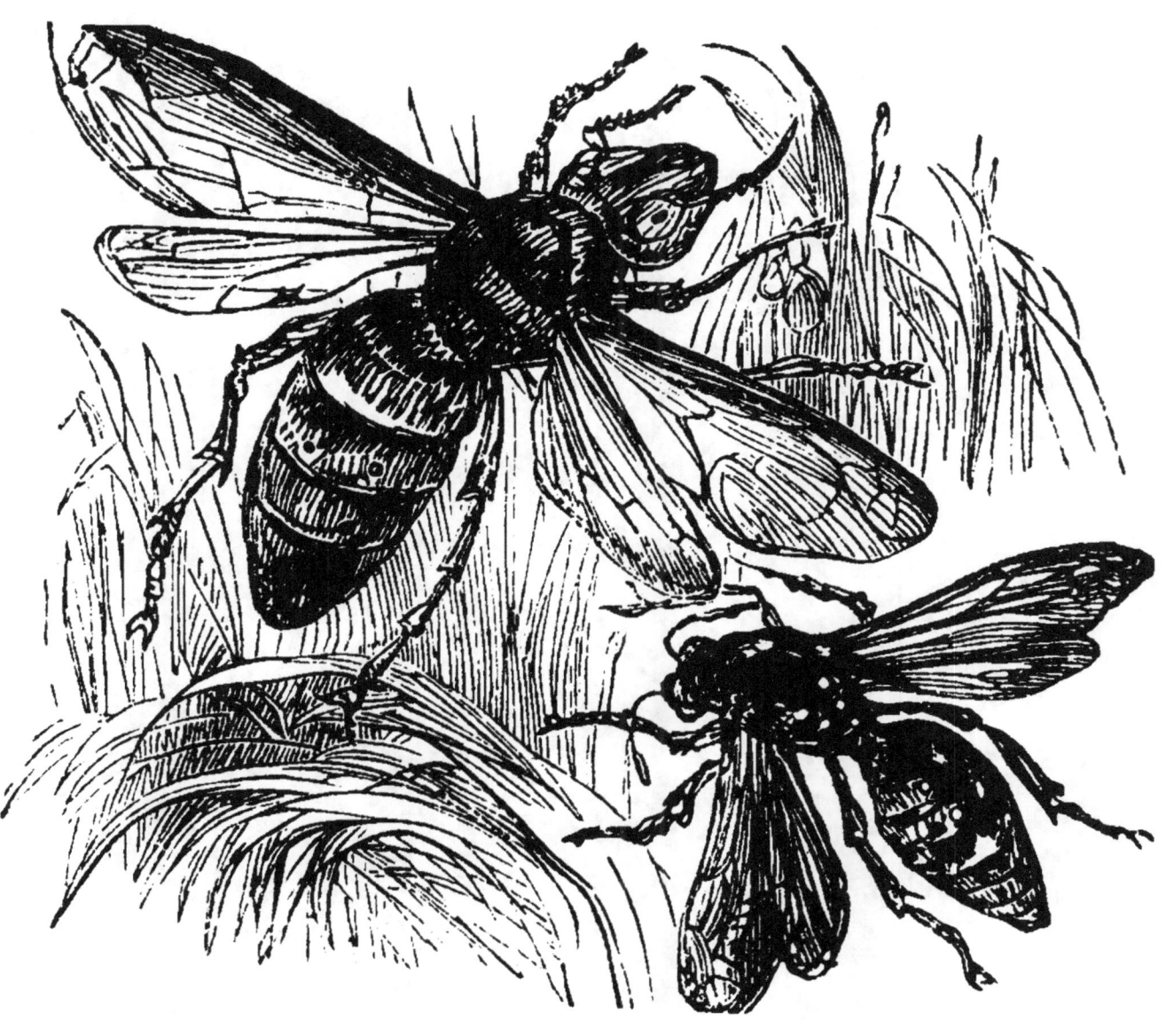

WASP SPECIES

There are thousands of wasp species, but they can be broadly categorized into two main types: social wasps and solitary wasps.

SOCIAL WASPS

Social Wasps live in colonies with a queen, workers, and males.

SOCIAL WASPS

Yellowjackets are black and yellow in appearance, smooth bodies.

SOCIAL WASPS

Yellowjackets are aggressive, especially in late summer. They nest in the ground or on structures.

SOCIAL WASPS

Paper wasps have long legs, slender bodies, brownish with yellow markings.

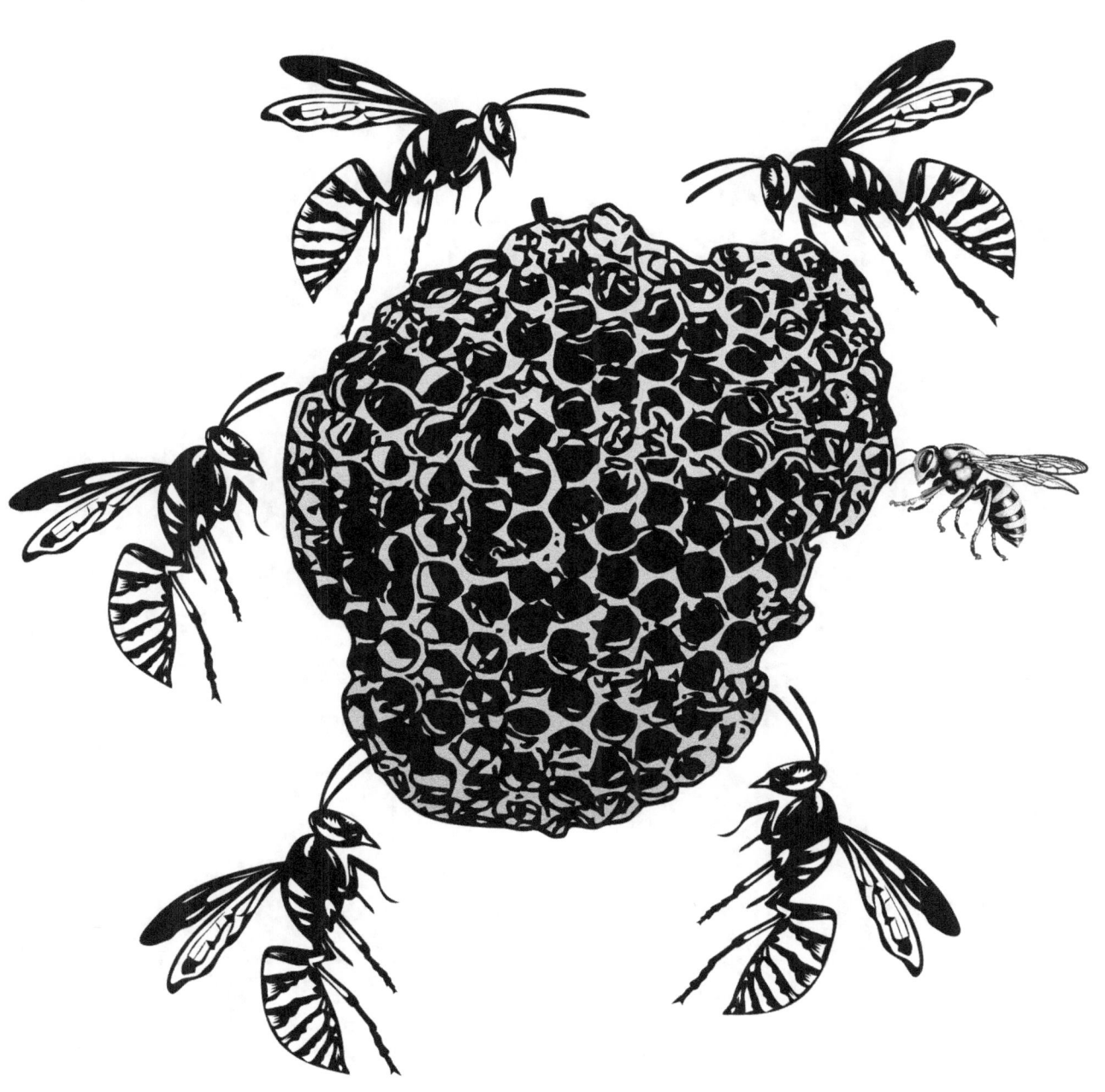

SOCIAL WASPS

Paper wasps make open, umbrella-shaped nests by chewing wood and mixing it with saliva, which gives the nests a paper-like texture.

SOCIAL WASPS

Hornets - a type of large yellowjacket - is larger, with white and black or brownish markings.

SOCIAL WASPS

Hornets can be aggressive. They build large aerial nests.

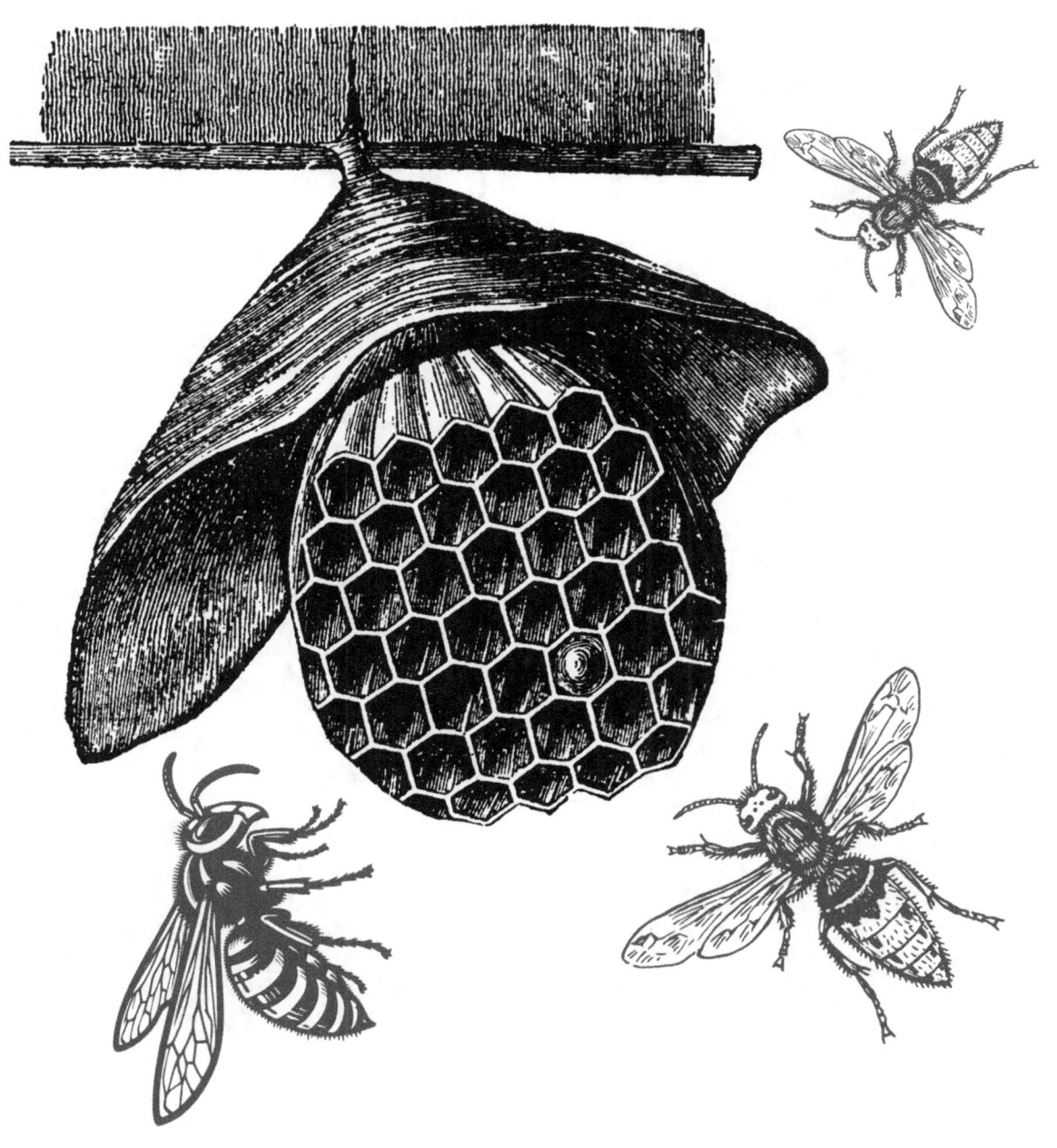

SOCIAL WASPS

Hornets eat insects and sweet substances.

SOCIAL WASPS

A distressed social wasp releases a pheromone that triggers nearby colony members to launch into a defensive, stinging attack.

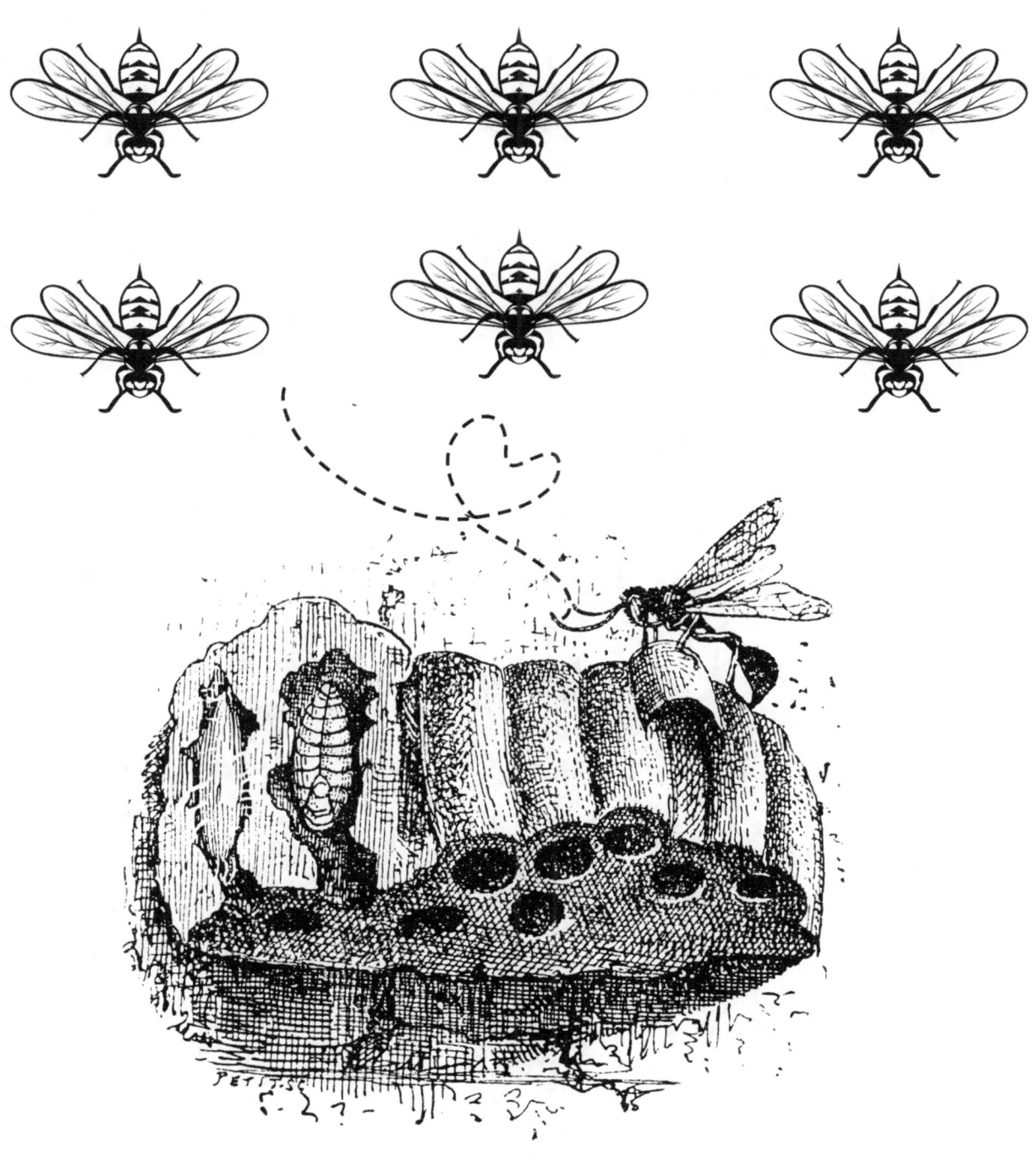

SOLITARY WASPS

Solitary wasps do not form colonies.   Each female builds her own nest.

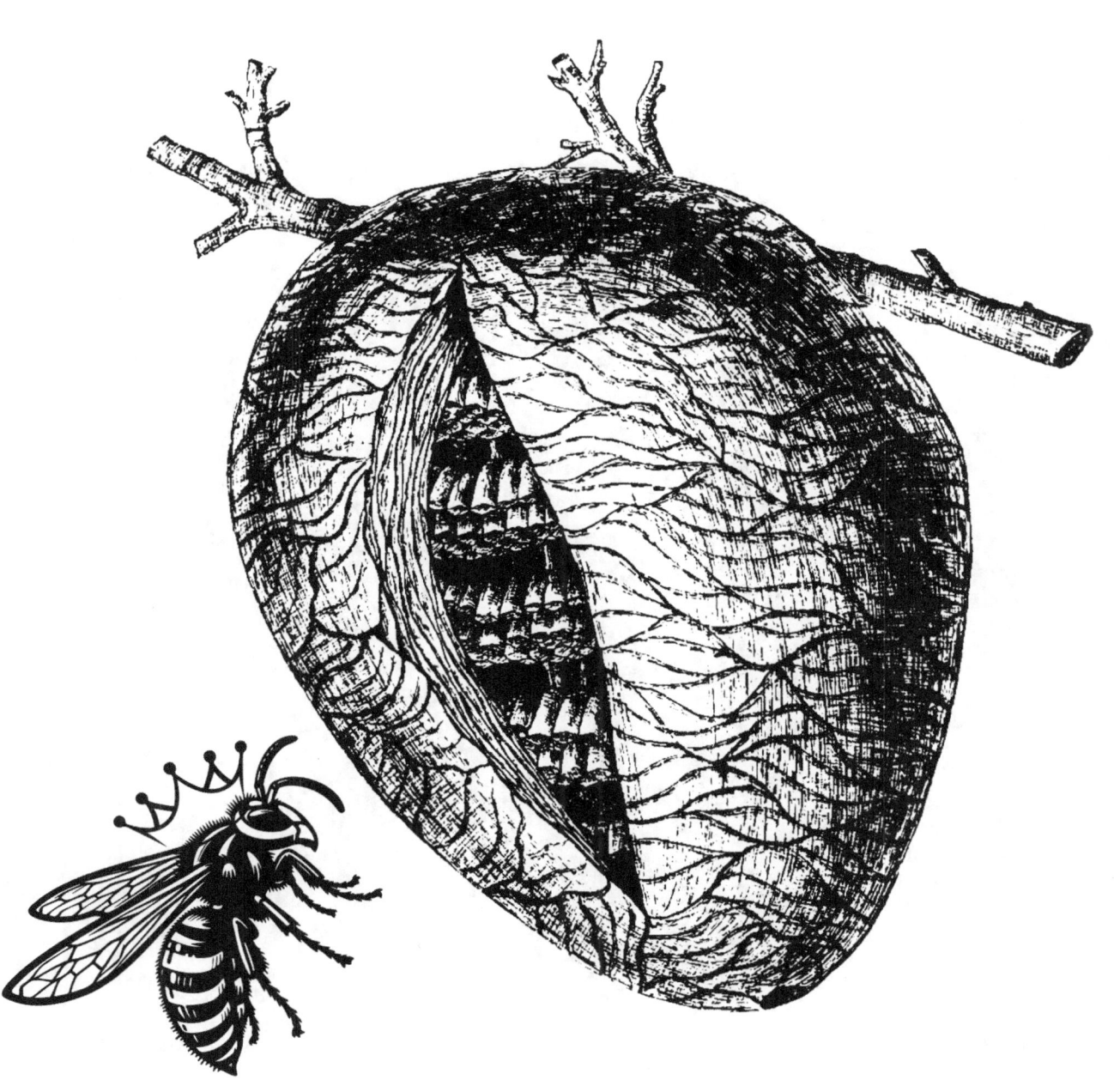

SOLITARY WASPS

Mud Daubers are long and slender with a narrow "waist". They are black or metallic blue. They build their nests out of mud.

MUD

SOLITARY WASPS

Mud Daubers are non-aggressive. They build mud tube nests to house their young.

SOLITARY WASPS

Mud Daubers diet include spiders, paralyzed and stored for larvae.

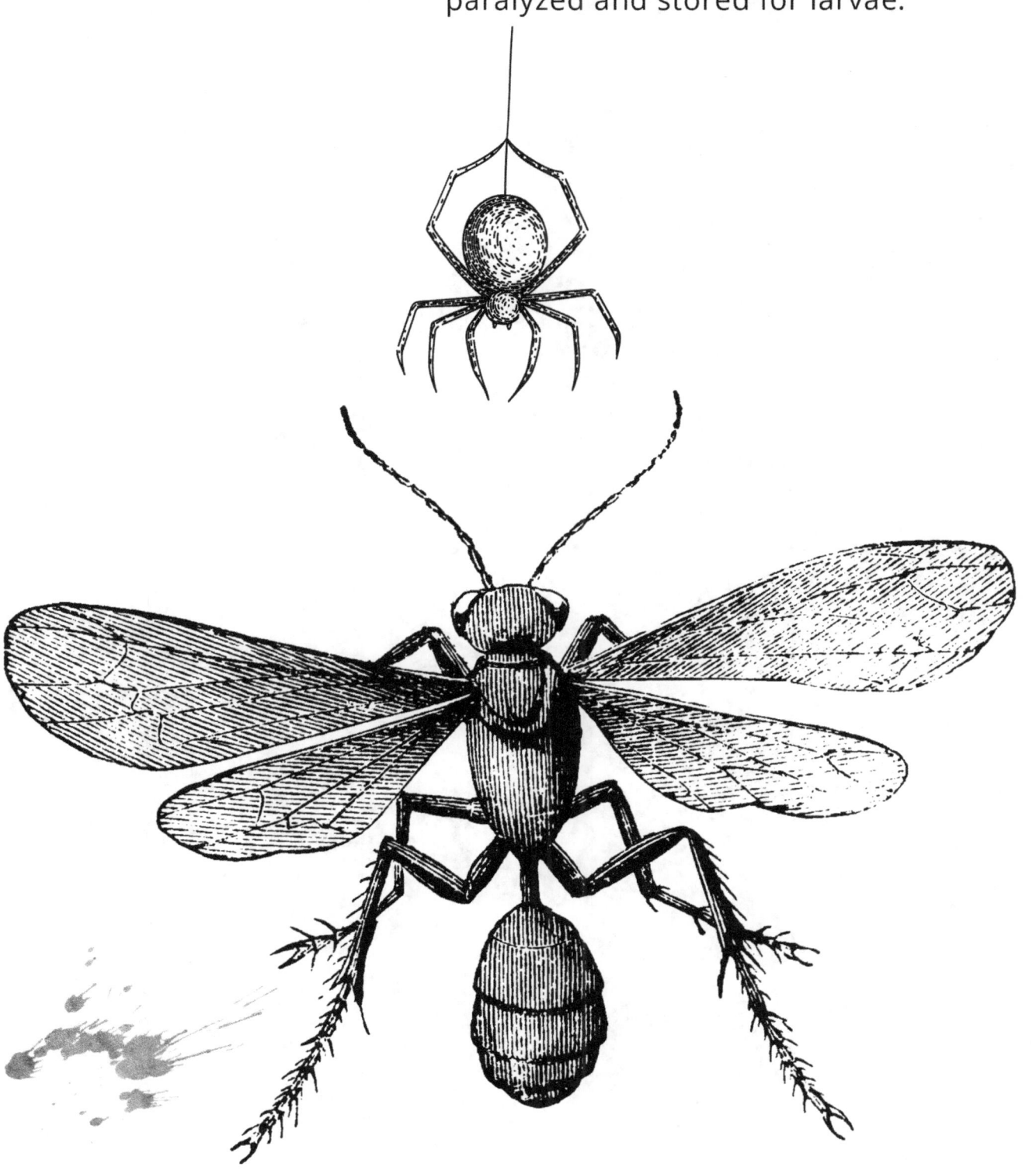

SOLITARY WASPS

Cicada Killers are large, black and yellow and resemble huge yellowjackets.

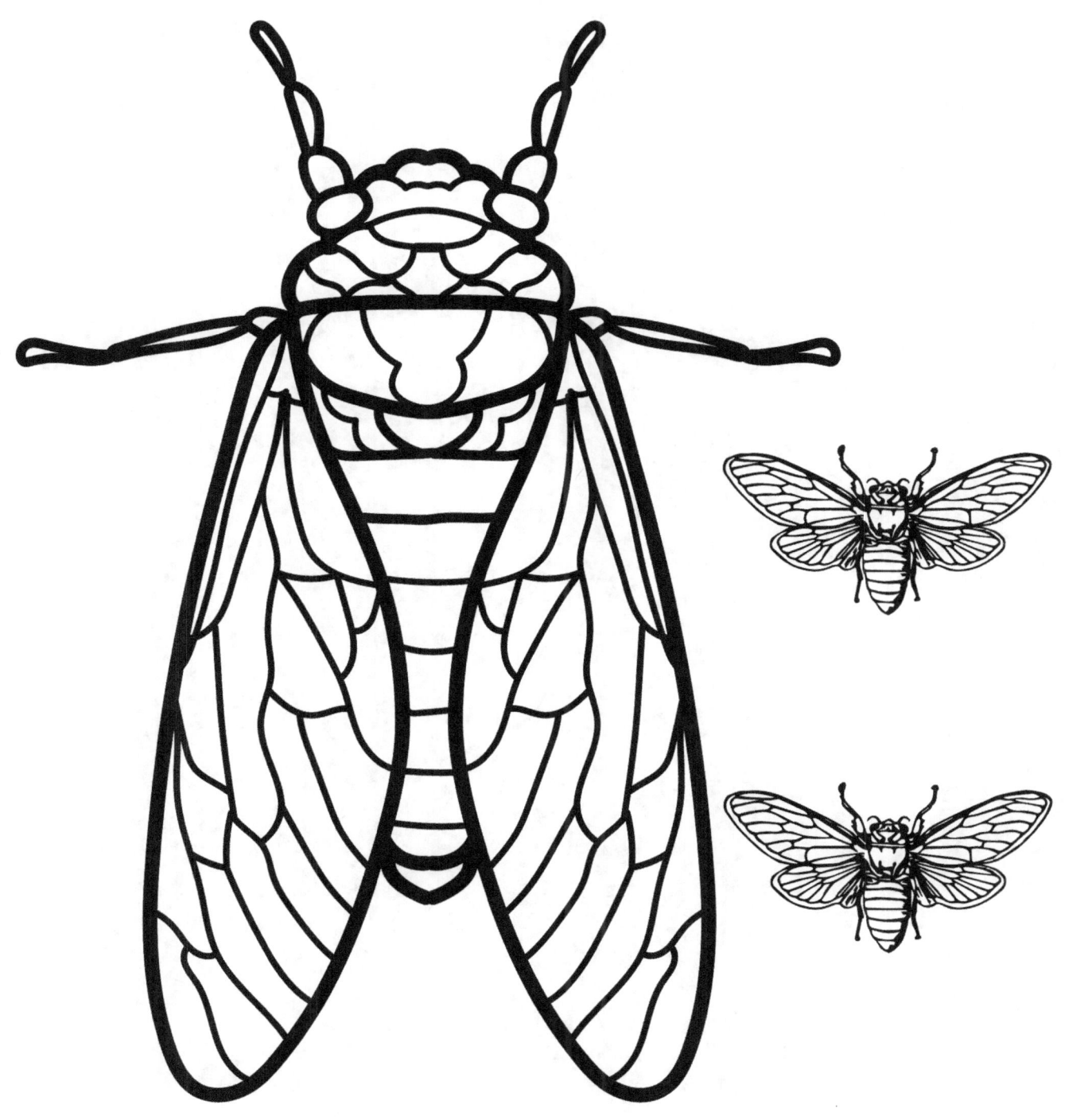

SOLITARY WASPS

Cicada Killers are solitary but can nest
in groups. Females dig burrows.

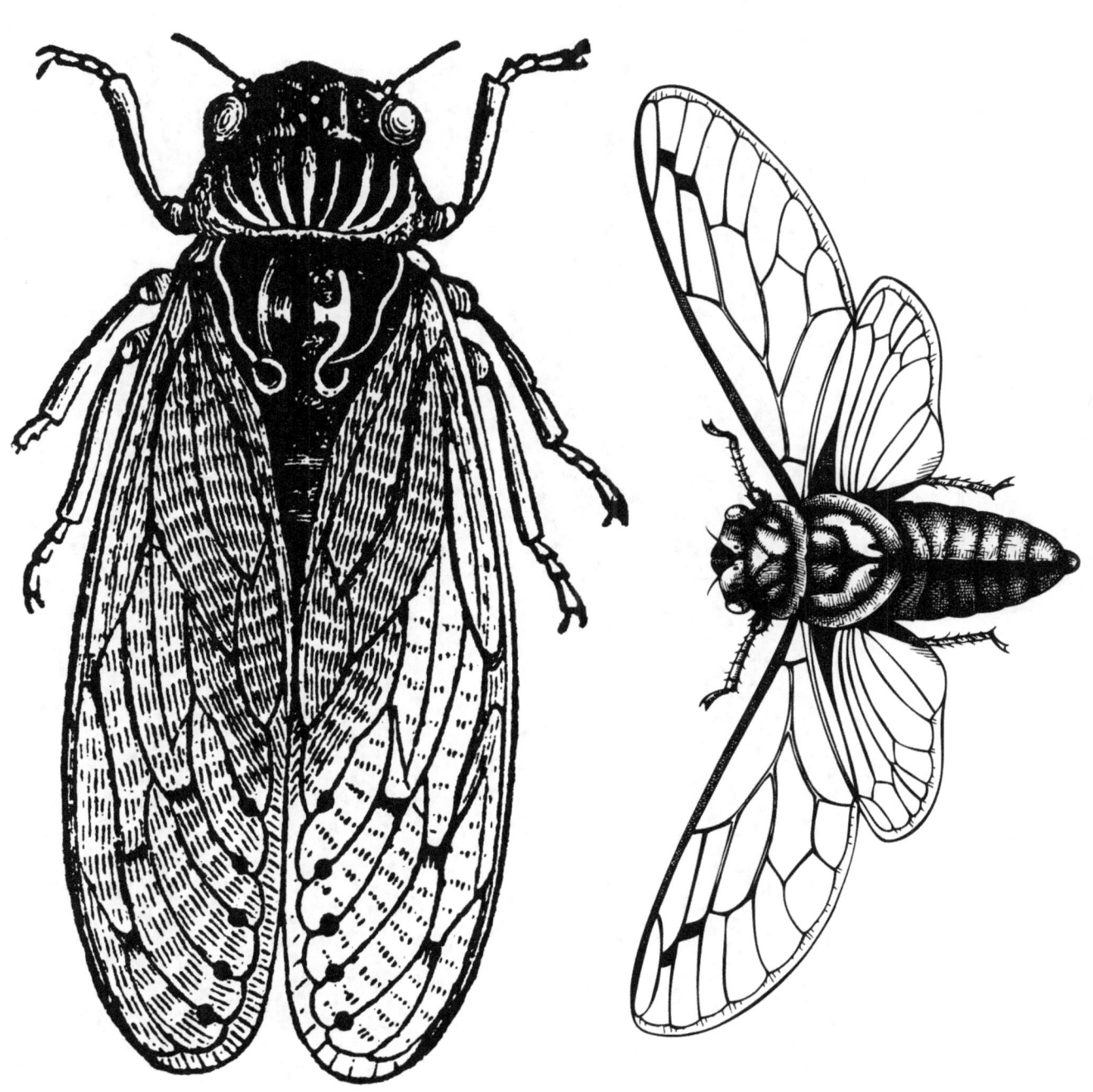

SOLITARY WASPS

The Cicada Killer's diet include cicadas, captured for larvae.

SOLITARY WASPS

Ichneumon wasps are often slender
with very long ovipositors.

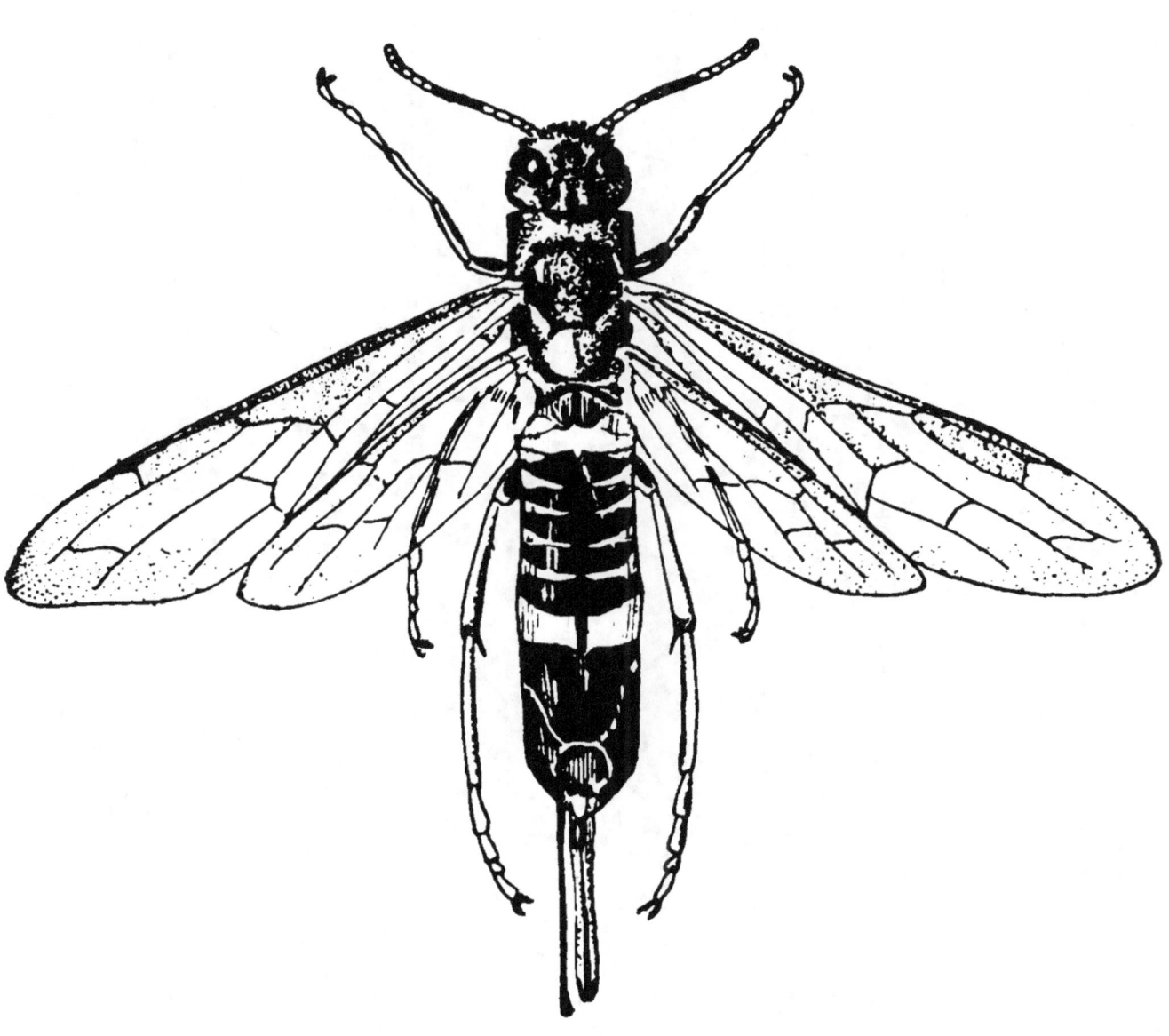

SOLITARY WASPS

Ichneumon Wasps are parasitic. They lay eggs in or on other insects.

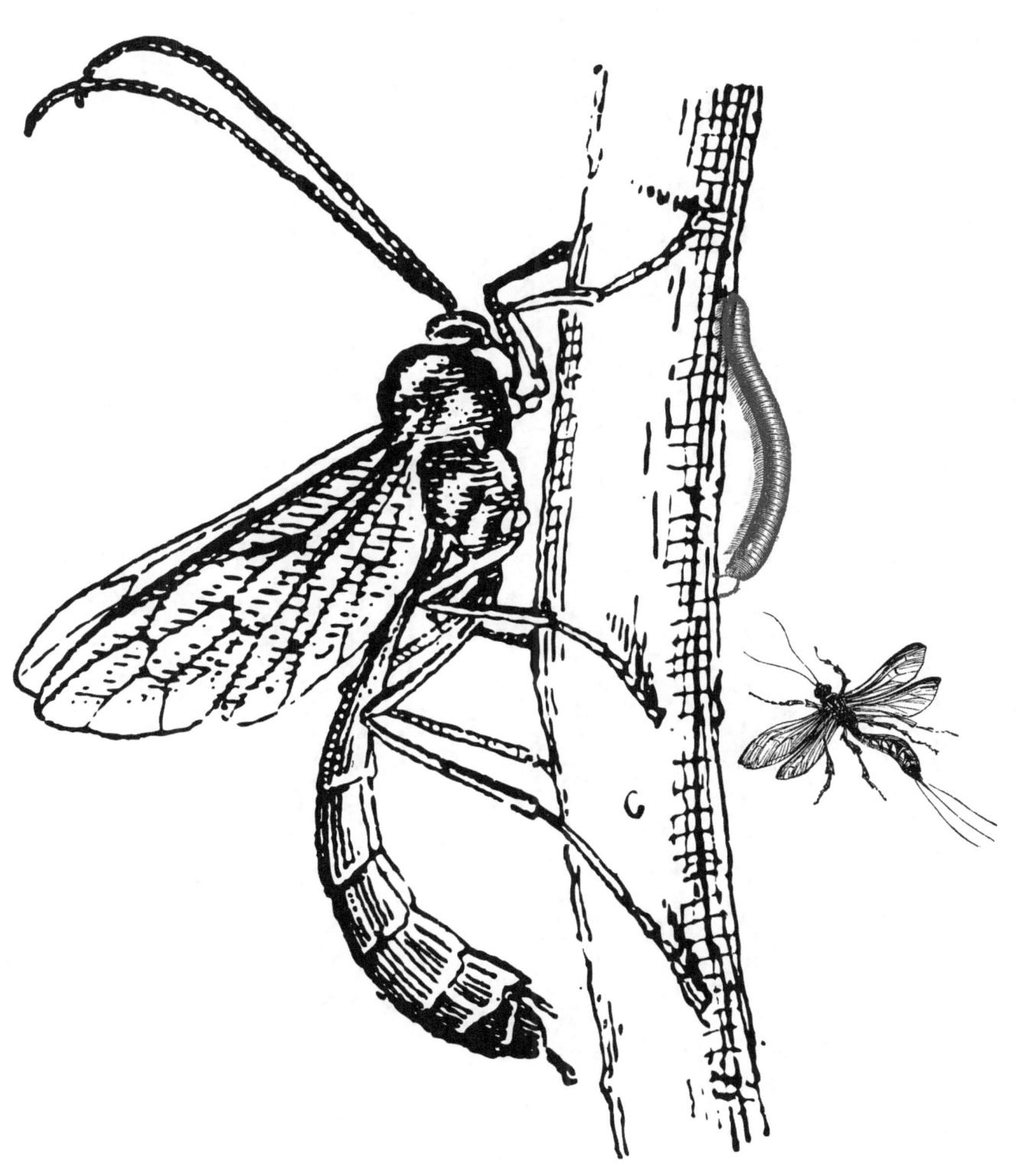

SOLITARY WASPS

Ichneumon Wasps diet include larvae
consume host insect from the inside.

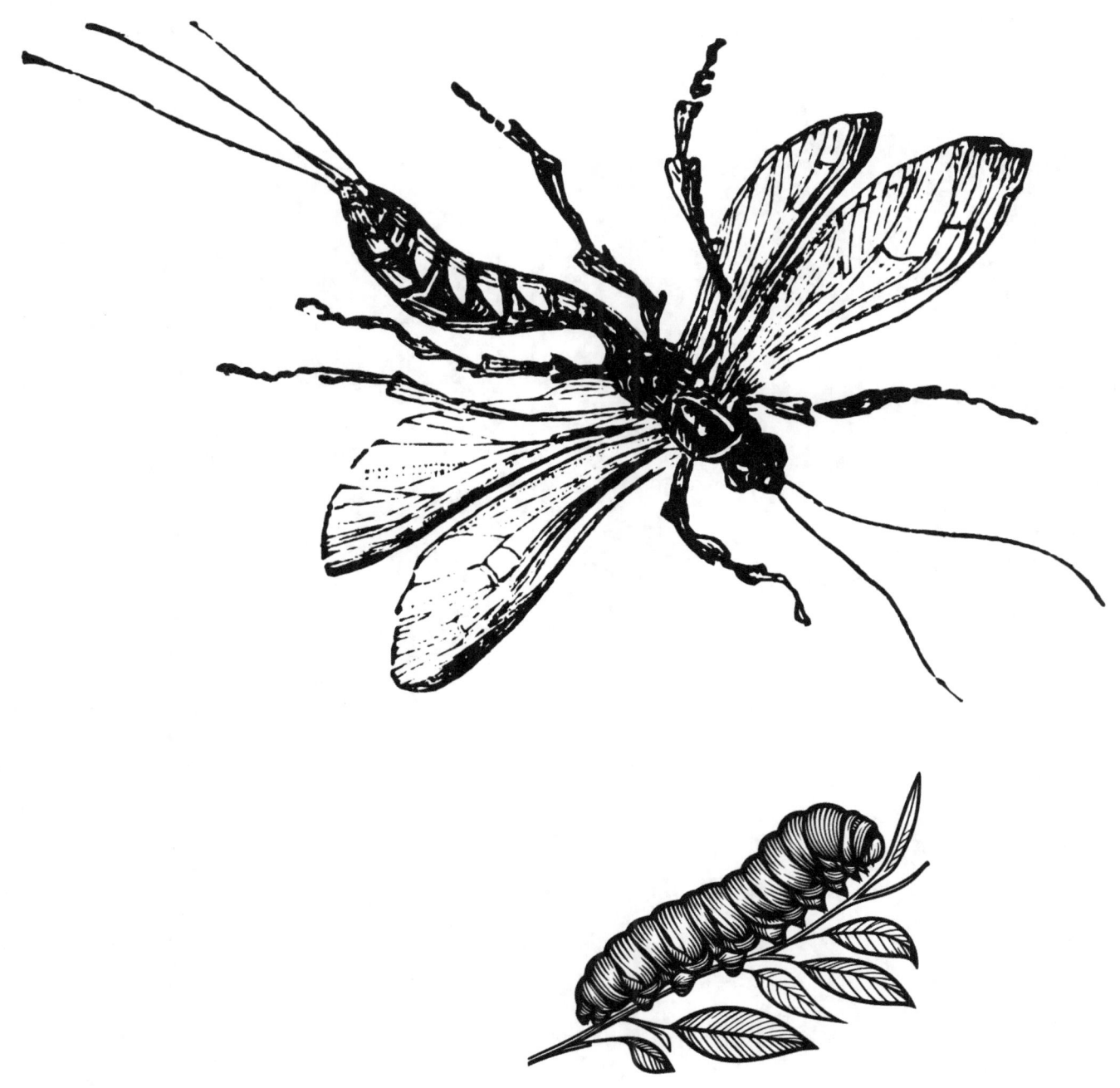

SOLITARY WASPS

Unlike bees, wasps are capable of stinging multiple times.

STINGER

Only female wasps can sting, and their stingers come from egg-laying organs.

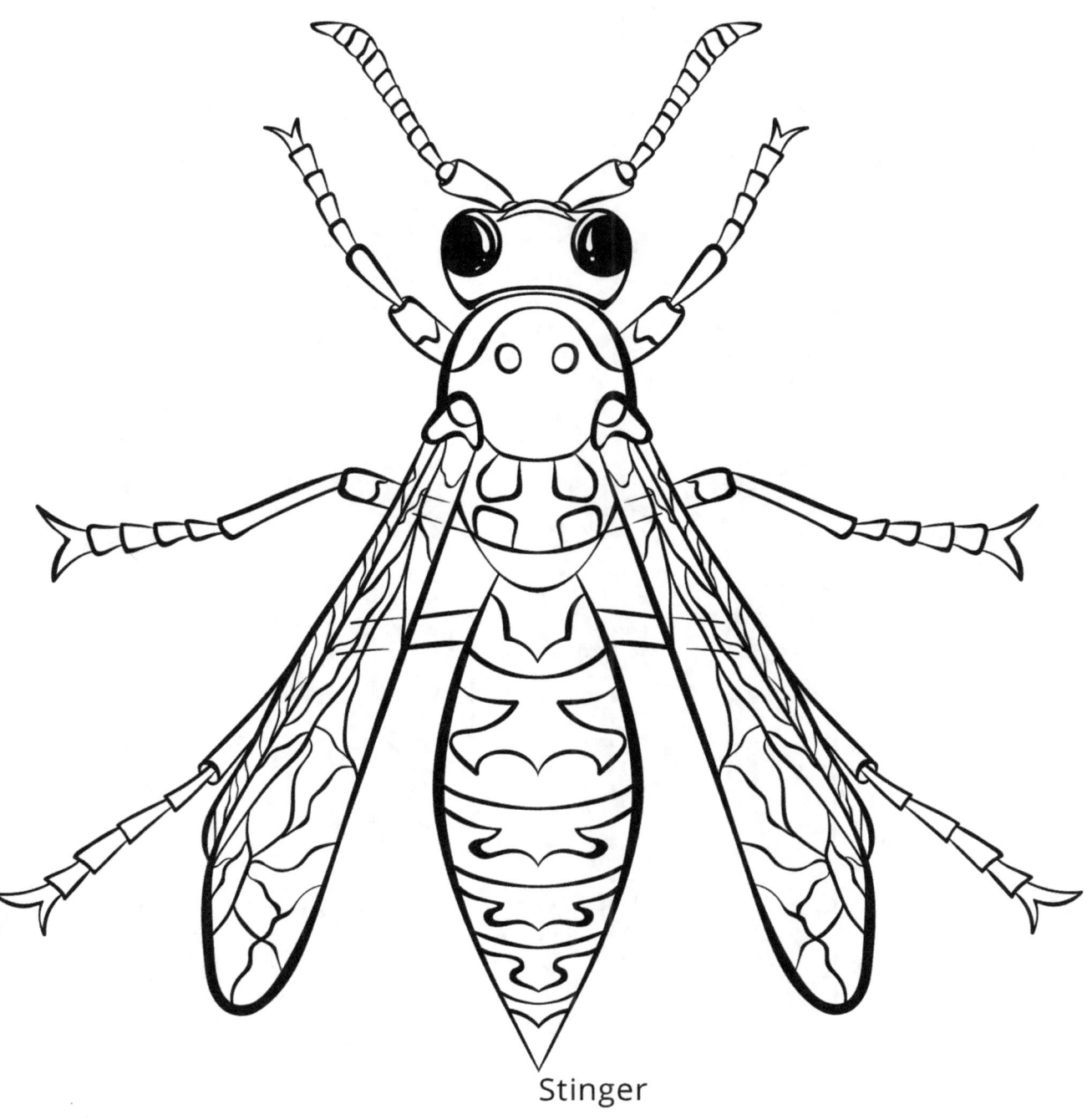

Stinger

STINGER

Most animals have developed a healthy respect for stinging wasps and tend to keep their distance.

STINGER

Creatures that accidentally stumble upon a wasp colony or dare to disturb a nest will quickly find themselves swarmed.

LOCAL ECOSYSTEMS

Despite their fearsome reputation, wasps are highly beneficial to humans.

LOCAL ECOSYSTEMS

Wasps are so good at controlling pests that farmers now use them to help protect crops.

LOCAL ECOSYSTEMS

Some insects in cities are helpful because they pollinate plants, eat pests, inspire science, and make fun places like butterfly gardens and bee farms.

LOCAL ECOSYSTEMS

Cities can make the most of good insects and reduce bad ones by building gardens, fixing buildings to keep pests out, and teaching people which bugs help and which ones don't.

PREDATORY

Some wasps are predatory and hunt other insects.

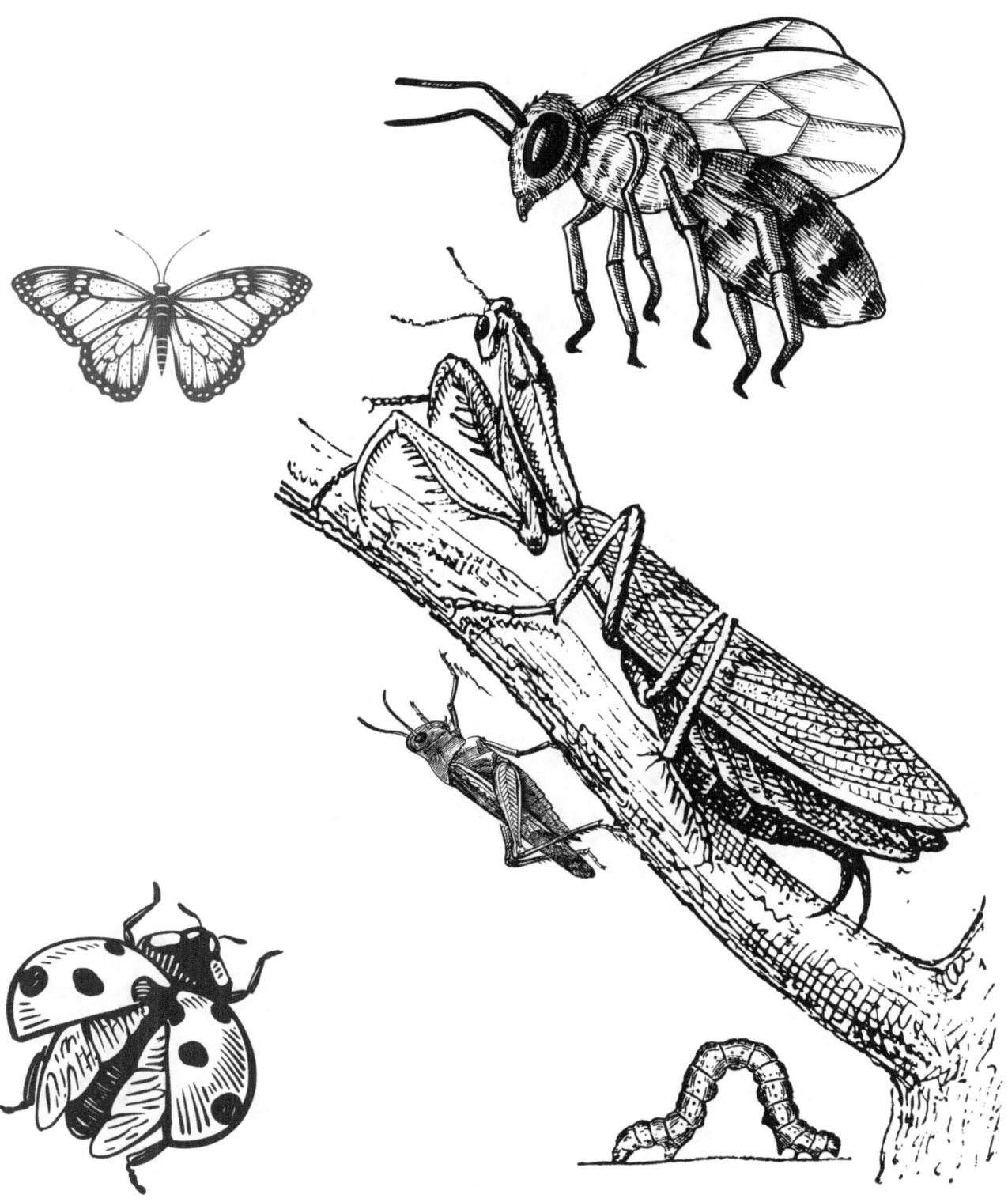

PARASITOIDS

Some wasps are parasitoids, laying eggs in or on other insects so their larvae can consume the host.

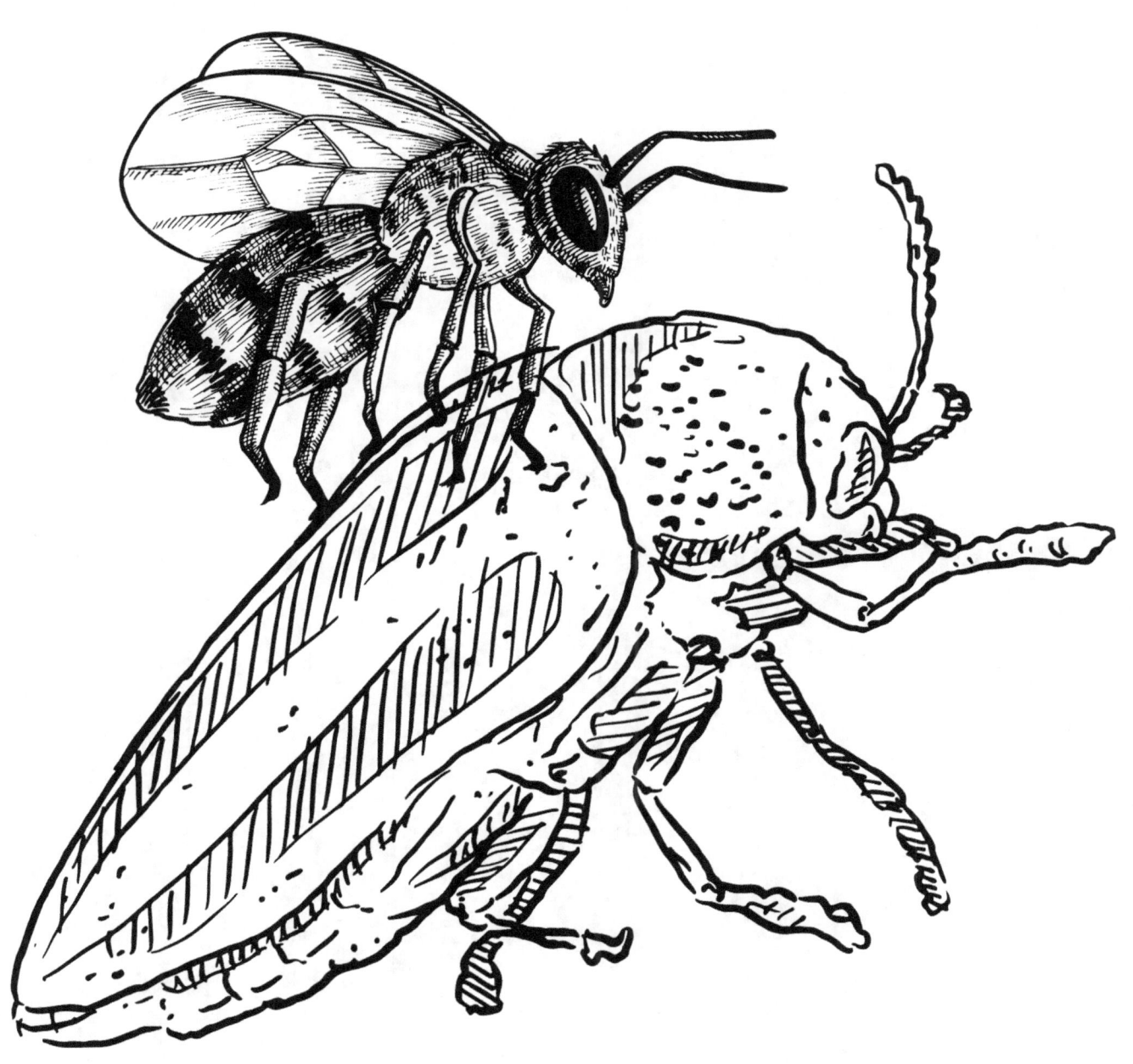

PREDATORY

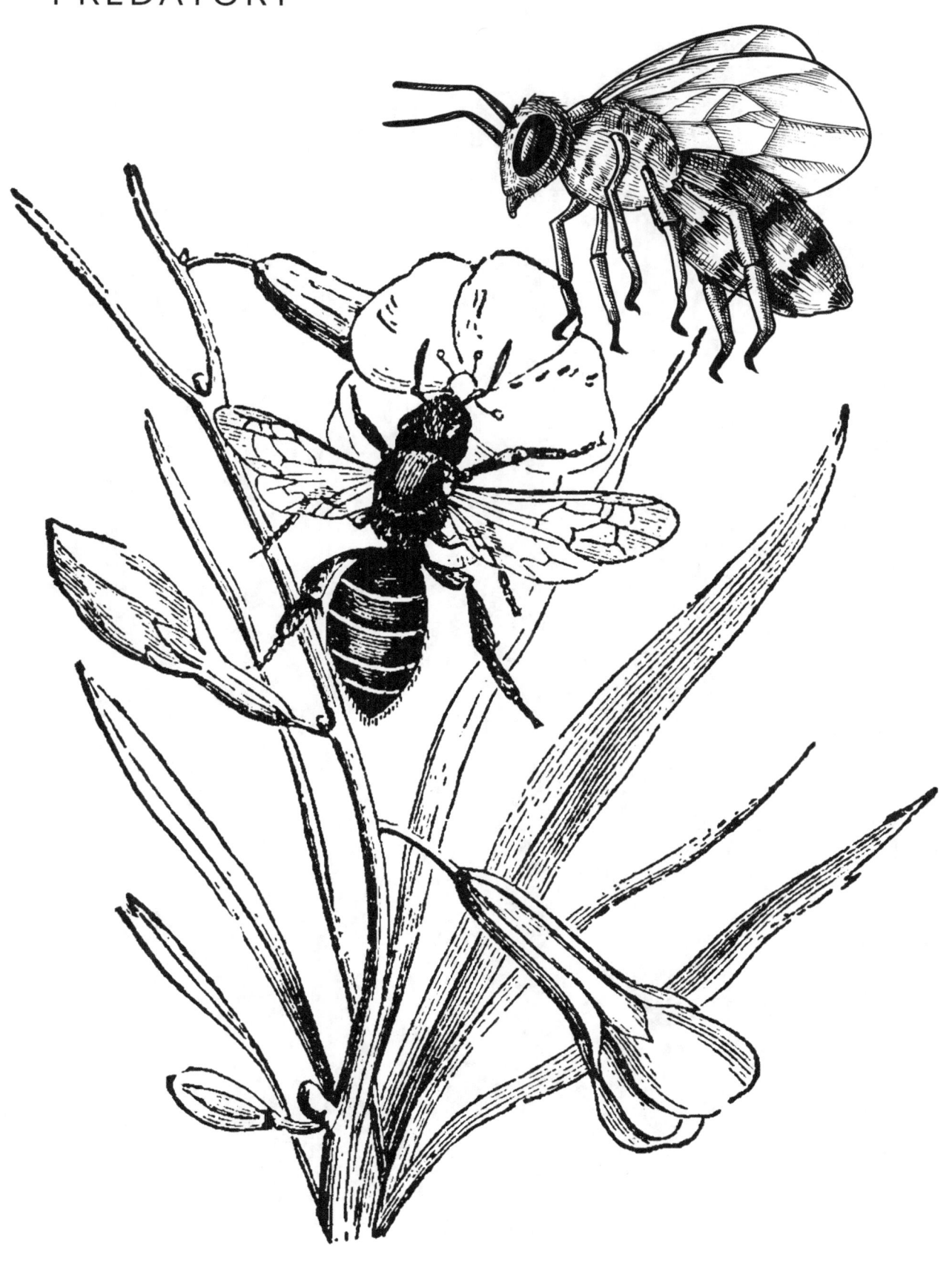

PARASITOIDS

PREDATORY

PARASITOIDS

POLLINATORS

Though wasps aren't as good at pollinating as bees, they still help flowers grow by spreading pollen when they visit for nectar.

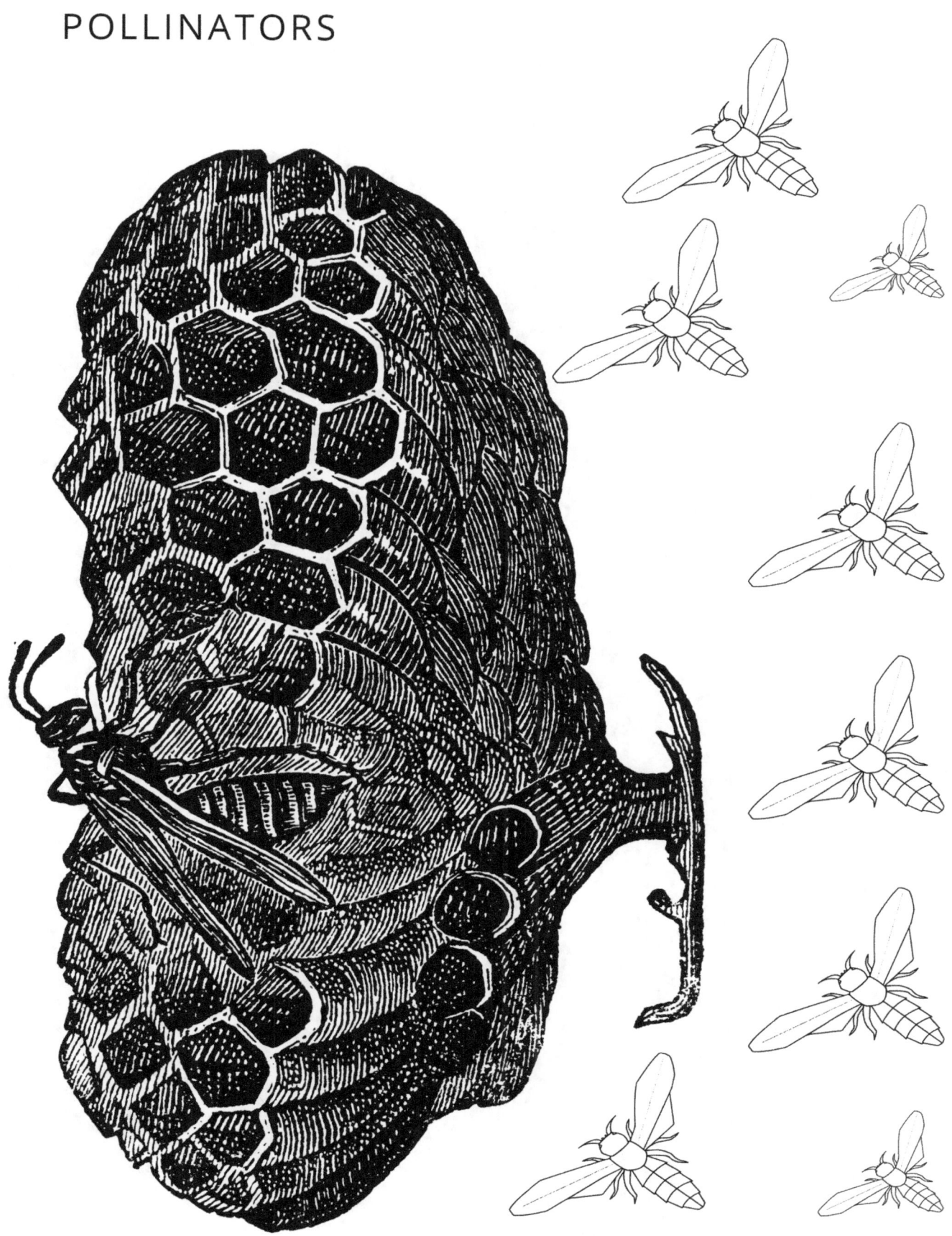

POLLINATORS

Wasps are important pollinators for some orchid species.

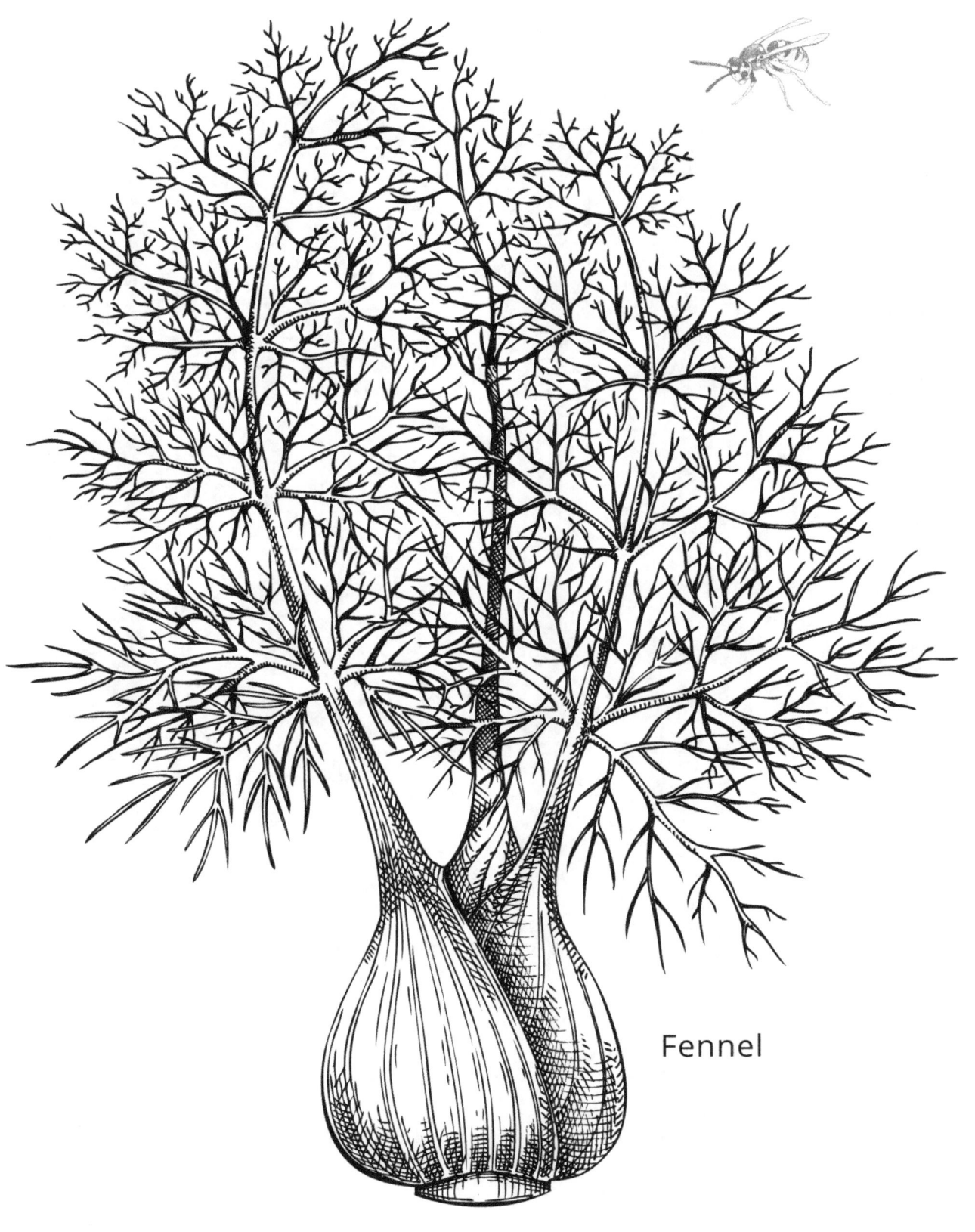

Fennel

PEST CONTROL

Wasps play a key role in protecting important crops like maize (corn) by acting as natural pest controllers.

PEST CONTROL

Wasps help stop sugarcane pests naturally, so plants grow better and farmers don't need to use as many chemicals.

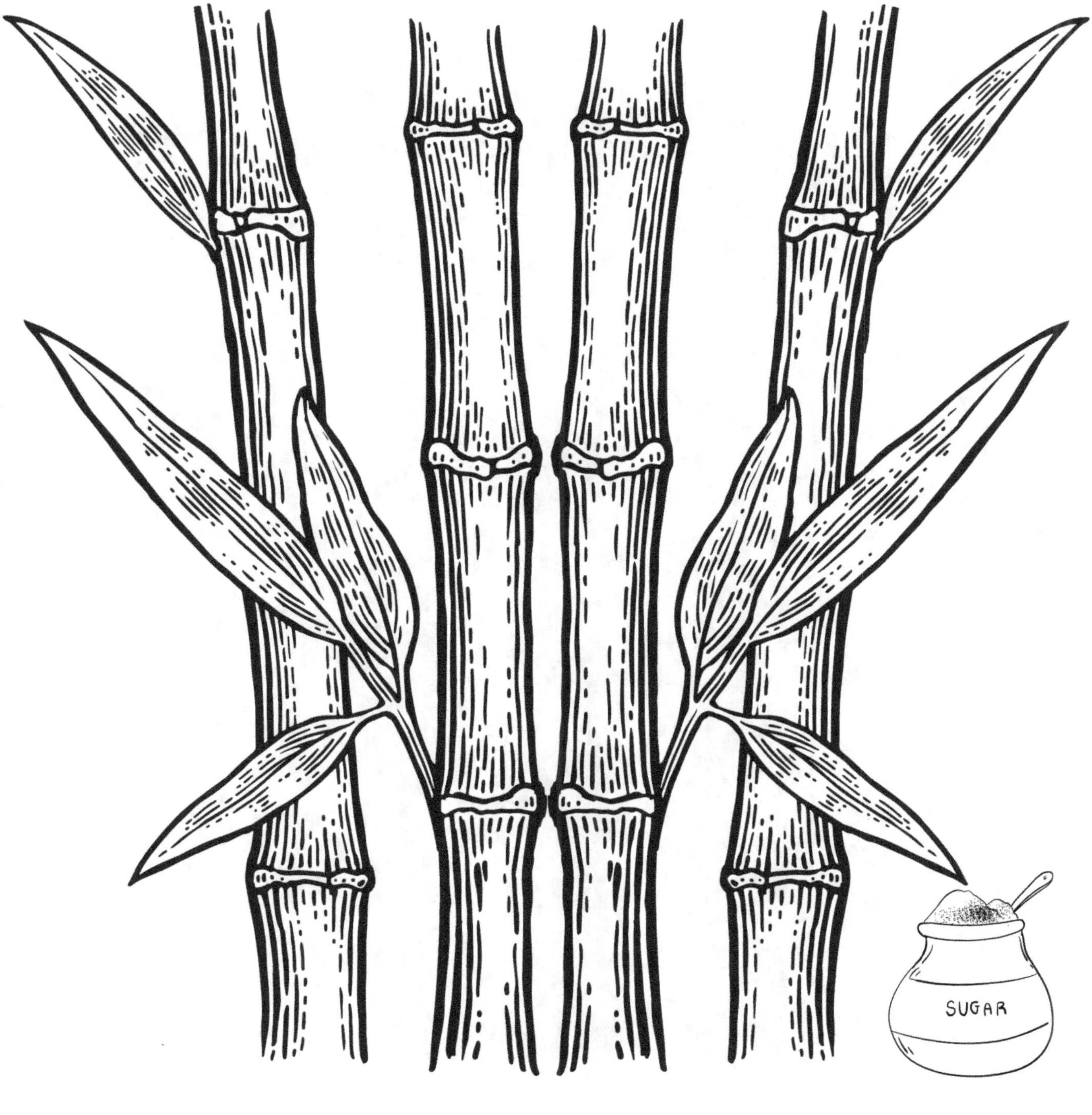

SUGAR

CROP PEST

Wasps can harm cherry crops, which can lead to farmers losing money.

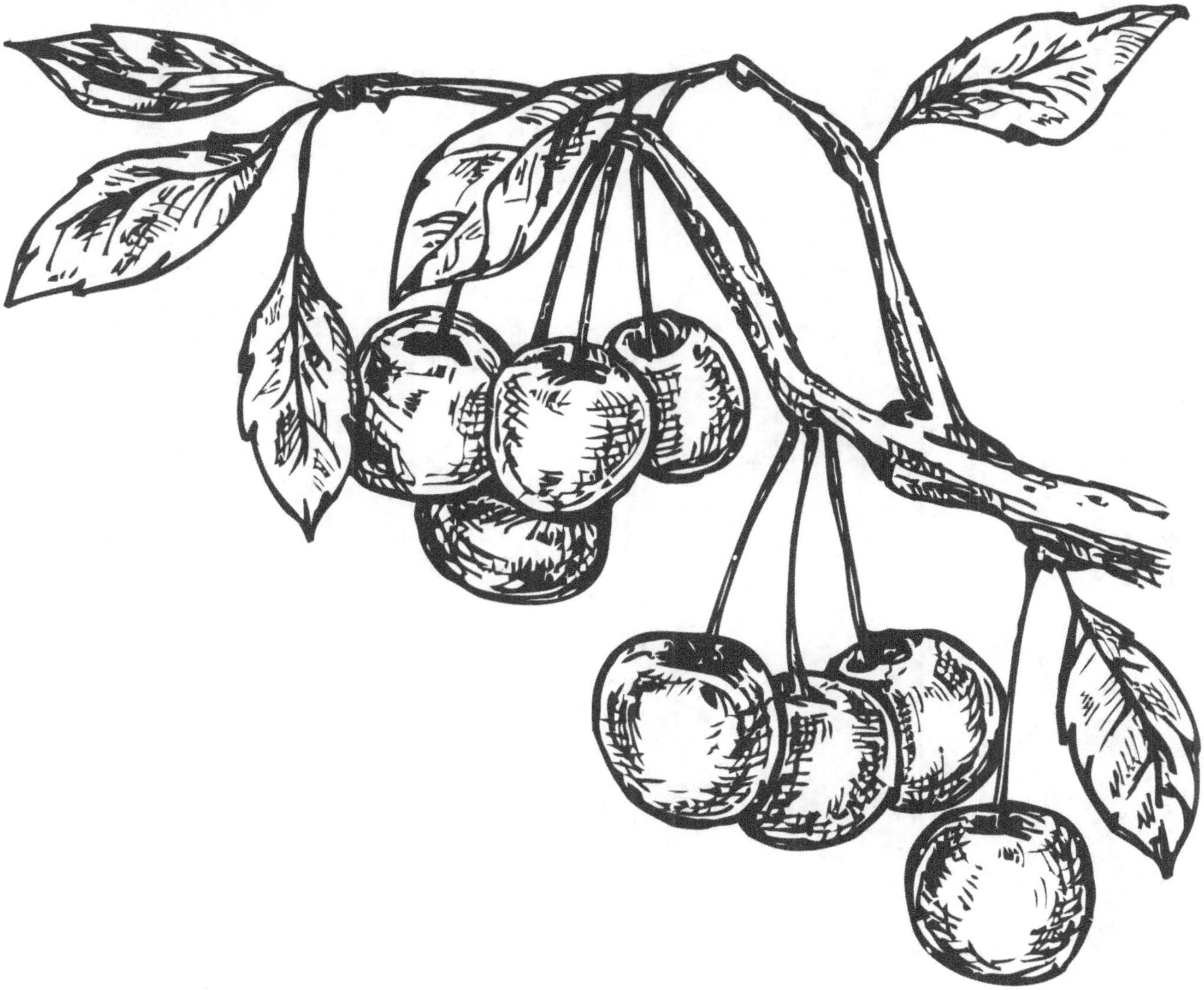

BENEFICIAL WASPS

Parsley and basil can bring helpful wasps that eat bugs that hurt plants.

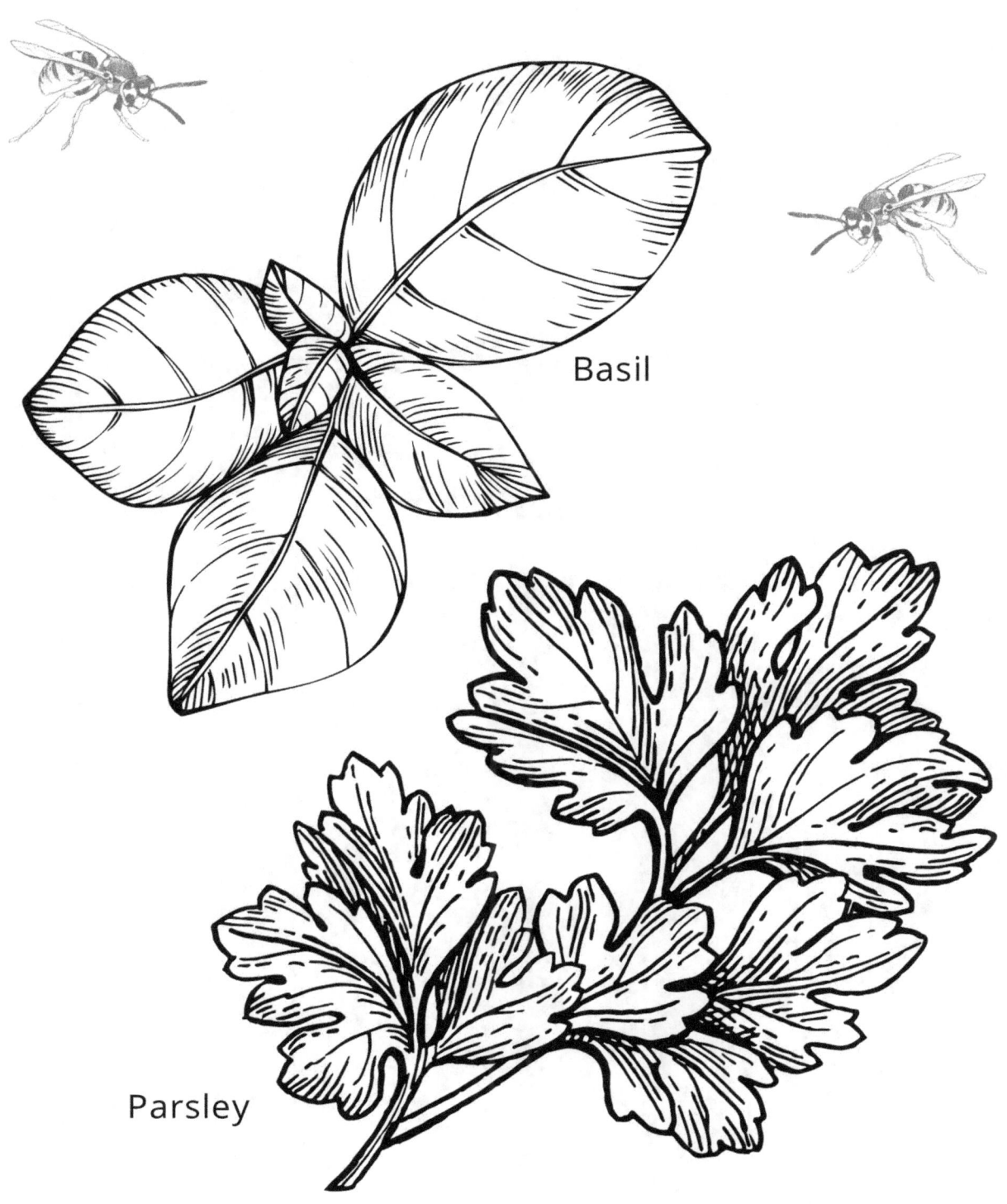

Basil

Parsley

BENEFICIAL WASPS

Marigolds bring in special wasps that help get rid of bugs that hurt chrysanthemum plants.

BENEFICIAL WASPS

Cilantro, dill, fennel, and buckwheat can bring helpful wasps and flies that fight off bad bugs.

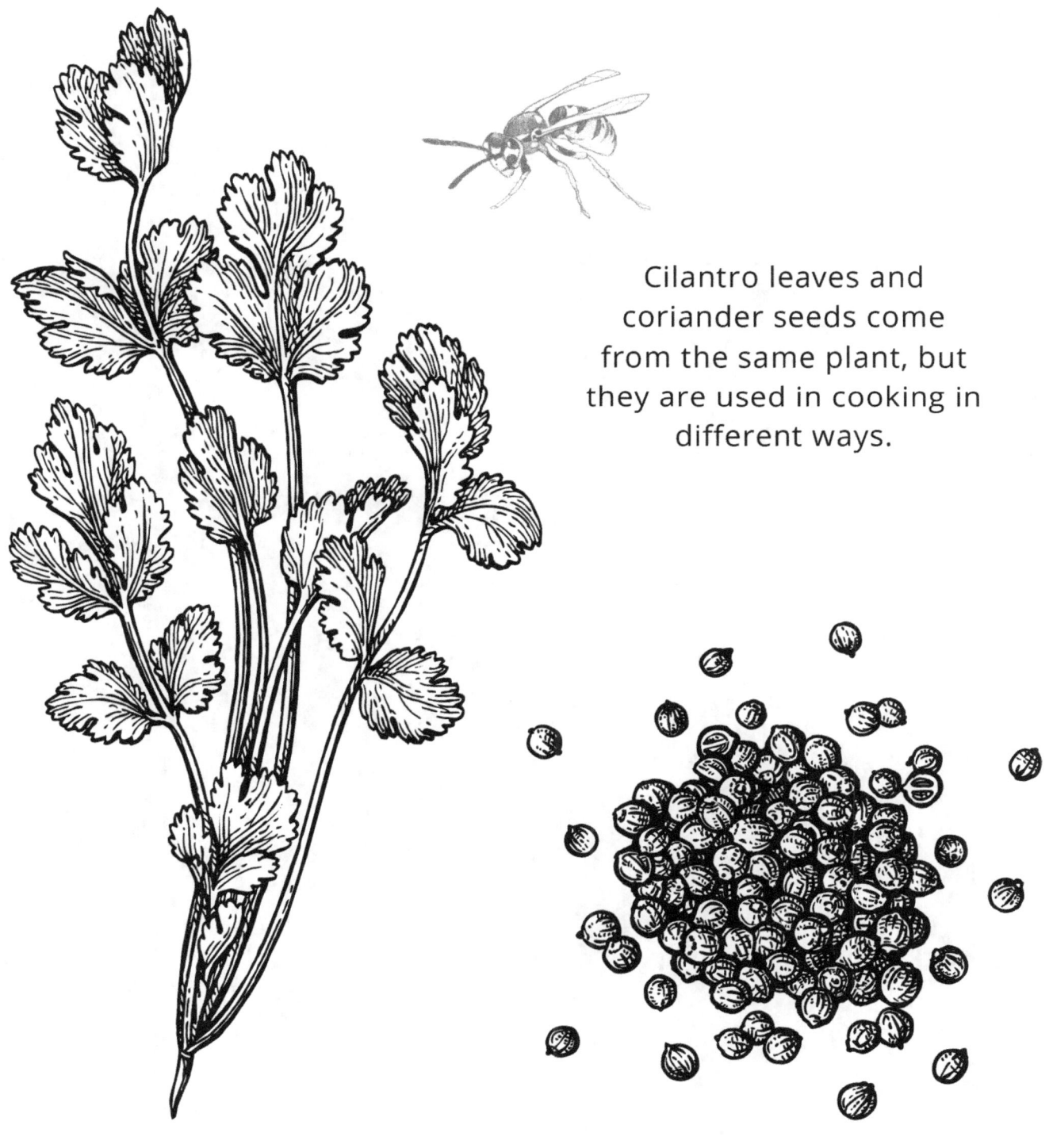

Cilantro leaves and coriander seeds come from the same plant, but they are used in cooking in different ways.

BENEFICIAL WASPS

Mint helps parasitic wasps by giving them food and a place to live.

BUG CATCHERS

Farmers and gardeners use parasitic wasps for biological pest control, as they attack and kill aphids, helping to reduce their numbers and protect plants.

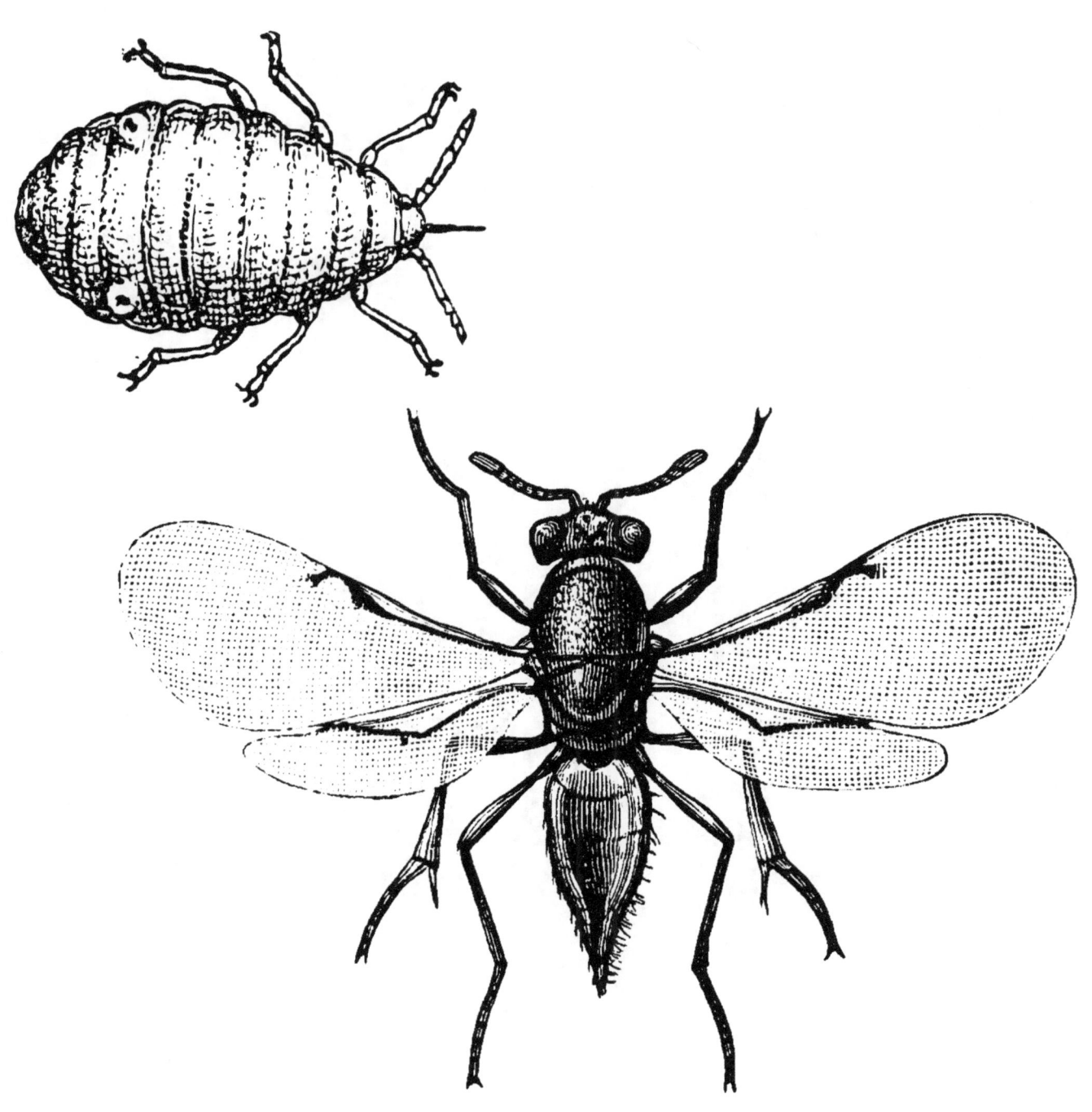

BUG CATCHERS

Some wasps enjoy visiting buckwheat plants because their flowers produce nectar, a sweet liquid that gives them energy.

BUG CATCHERS

Wasps catch grasshoppers by stinging them to make them sleepy and then take them back to feed their babies.

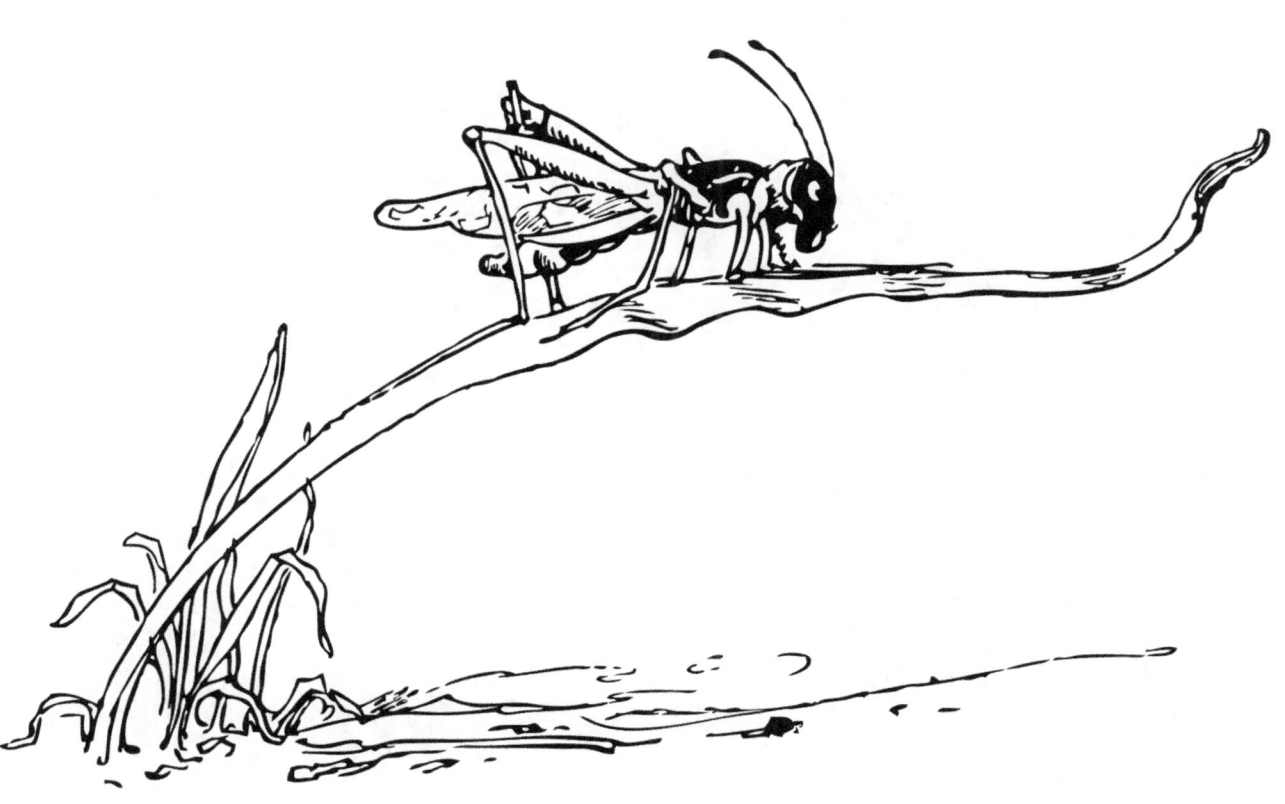

BUG CATCHERS

Wasps catch beetles by stinging them to make them stop moving, then take them home to feed their babies.

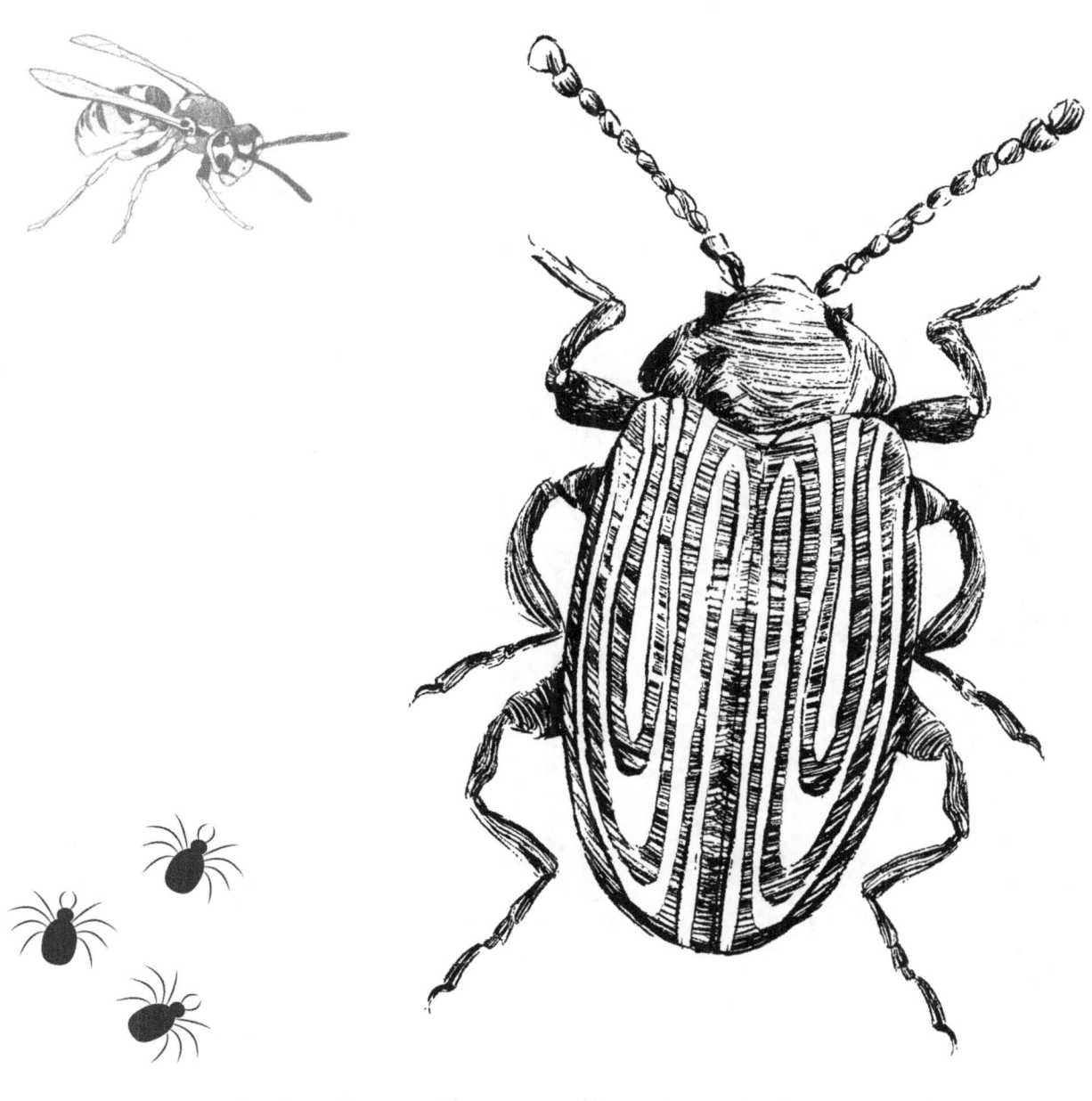

BUG CATCHERS

Wasps catch moths by stinging them to make them sleepy, then bring them home to feed their babies.

NEST BUILDERS

Wasps make their nests in trees, under porches, or sometimes in the ground.

Oak Tree

NEST BUILDERS

Be careful near apple trees with ripe fruit, especially in late summer and fall - wasps are busy and like sweet fruit.

NEST BUILDERS

Wasps build nests in maple trees using chewed-up wood, which they turn into a gray, paper-like material!

NEST BUILDERS

Wasps are often found around pear trees because they love the sweet, juicy fruit.

NEST BUILDERS

Pine trees attract insects like caterpillars and beetles, which wasps hunt to feed their babies.

Pine Tree

NEST BUILDERS

Cedar trees have sturdy branches and dense foliage, offering a safe, sheltered place to build nests.

FOREST ECOSYSTEMS

Forest ecosystems give wasps a calm, natural home, far from people and buildings.

FOREST ECOSYSTEMS

Wasps play an important role in forest ecosystems by helping keep nature in balance.

FOREST ECOSYSTEMS

Some wasps in forests help trees by controlling pests and pollinating wildflowers!

FOREST ECOSYSTEMS

Wasps help forests by eating bugs, spreading flower pollen, and being food for other animals!

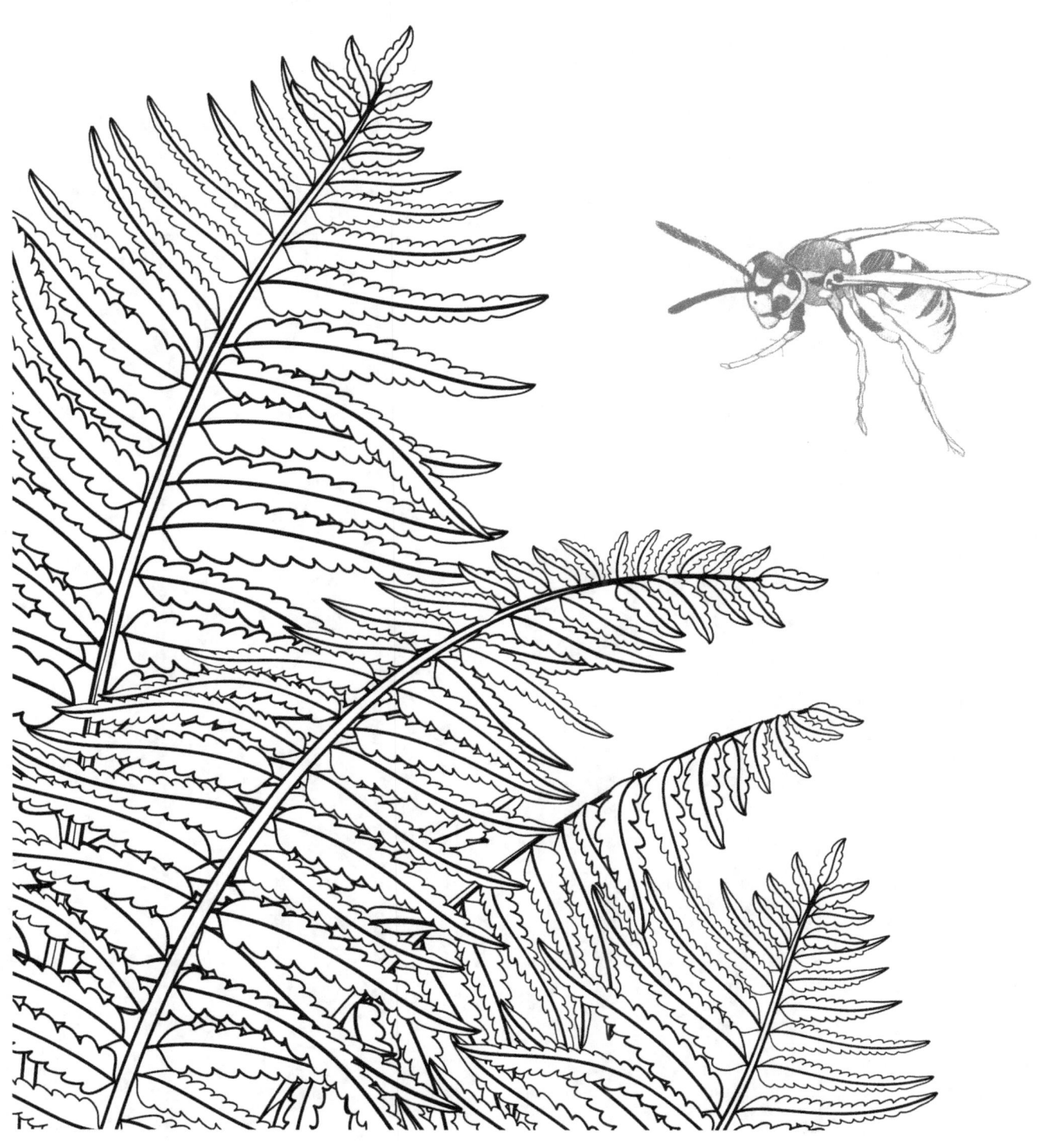

AGRICULTURE ECOSYSTEMS

An agricultural ecosystem is a place like a farm where plants, animals, soil, water, and people all help each other to grow food.

AGRICULTURE ECOSYSTEMS

Wasps help crop rotation by eating bad bugs, keeping plants safe as farmers grow different crops in the same field..

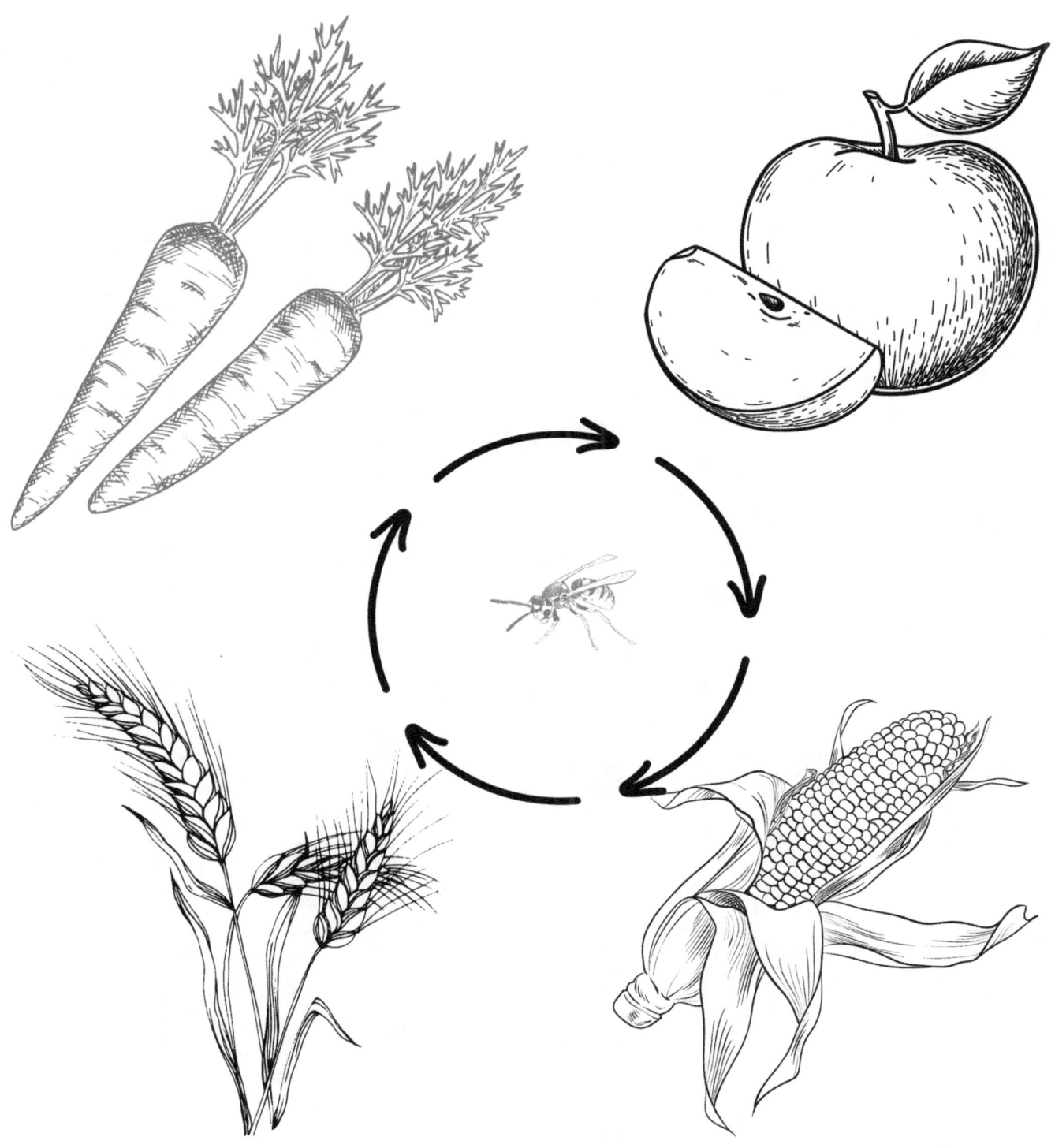

AGRICULTURE ECOSYSTEMS

Wasps are very helpful in farming and agricultural areas because they act as natural pest control and sometimes help with pollination.

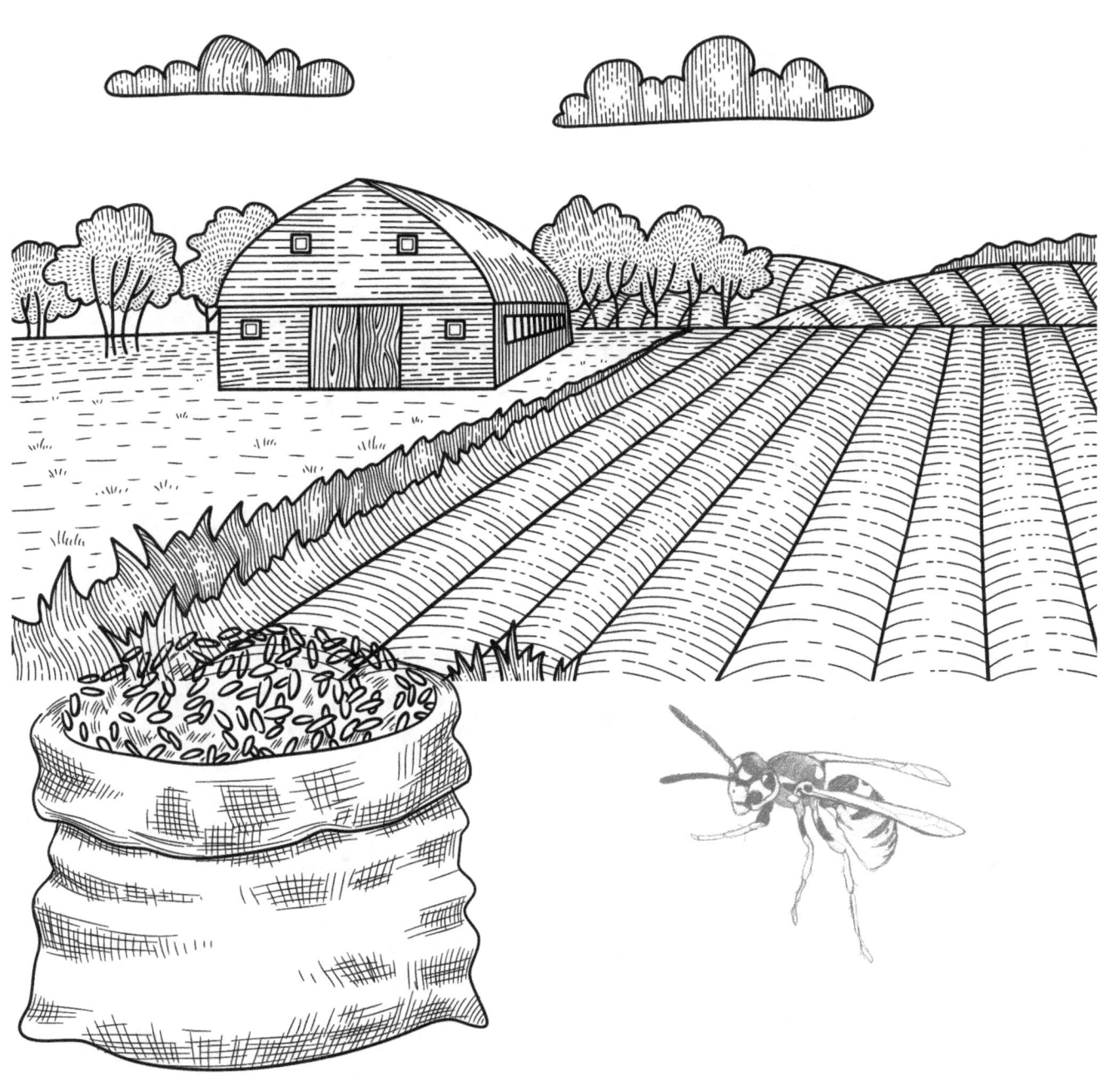

AGRICULTURE ECOSYSTEMS

By hunting harmful insects, wasps reduce the need for chemical pesticides, which can hurt the environment and other helpful bugs like bees.

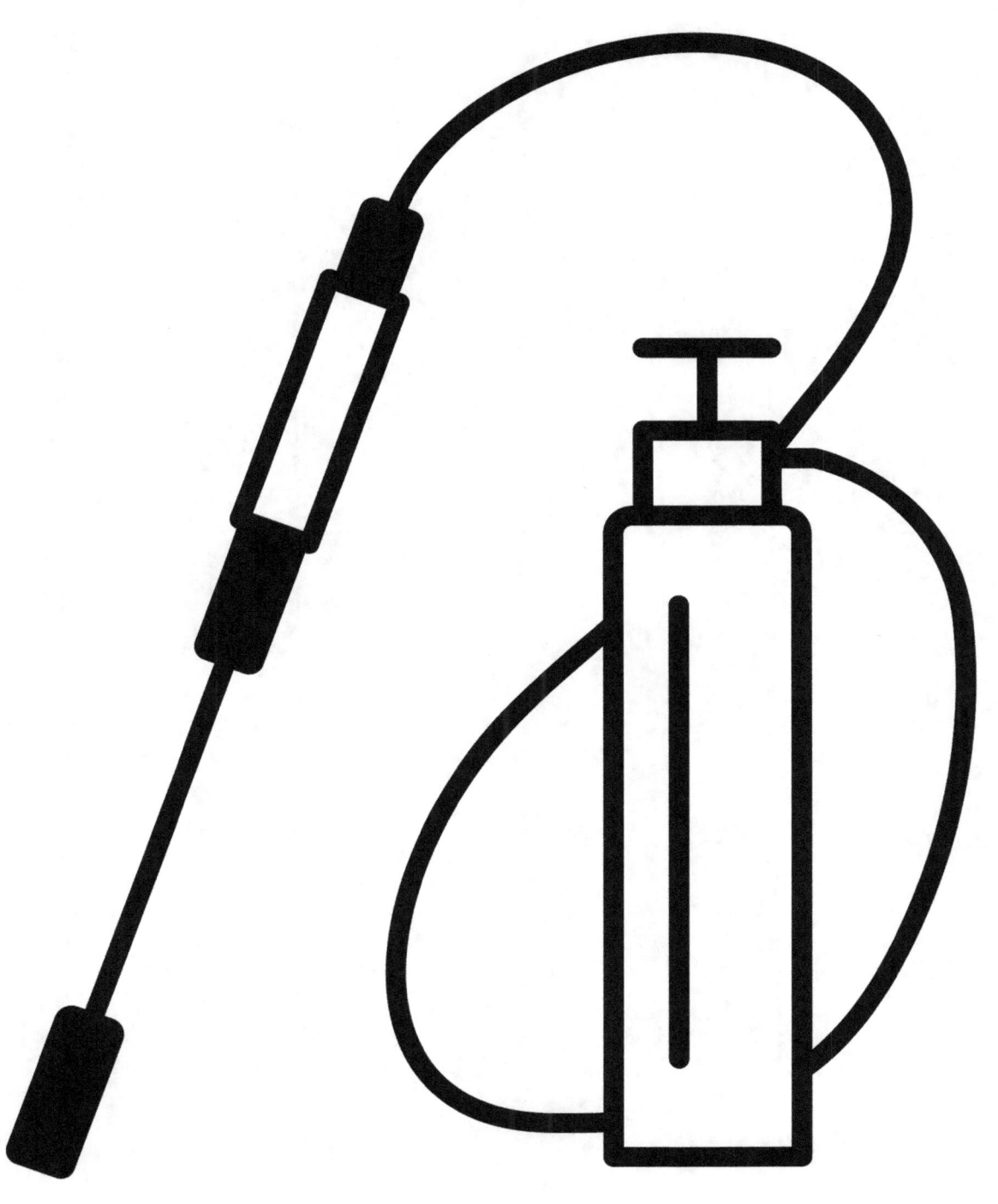

AGRICULTURE ECOSYSTEMS

Wasps are part of the food web. They help keep bug populations balanced and also serve as food for birds and other animals.

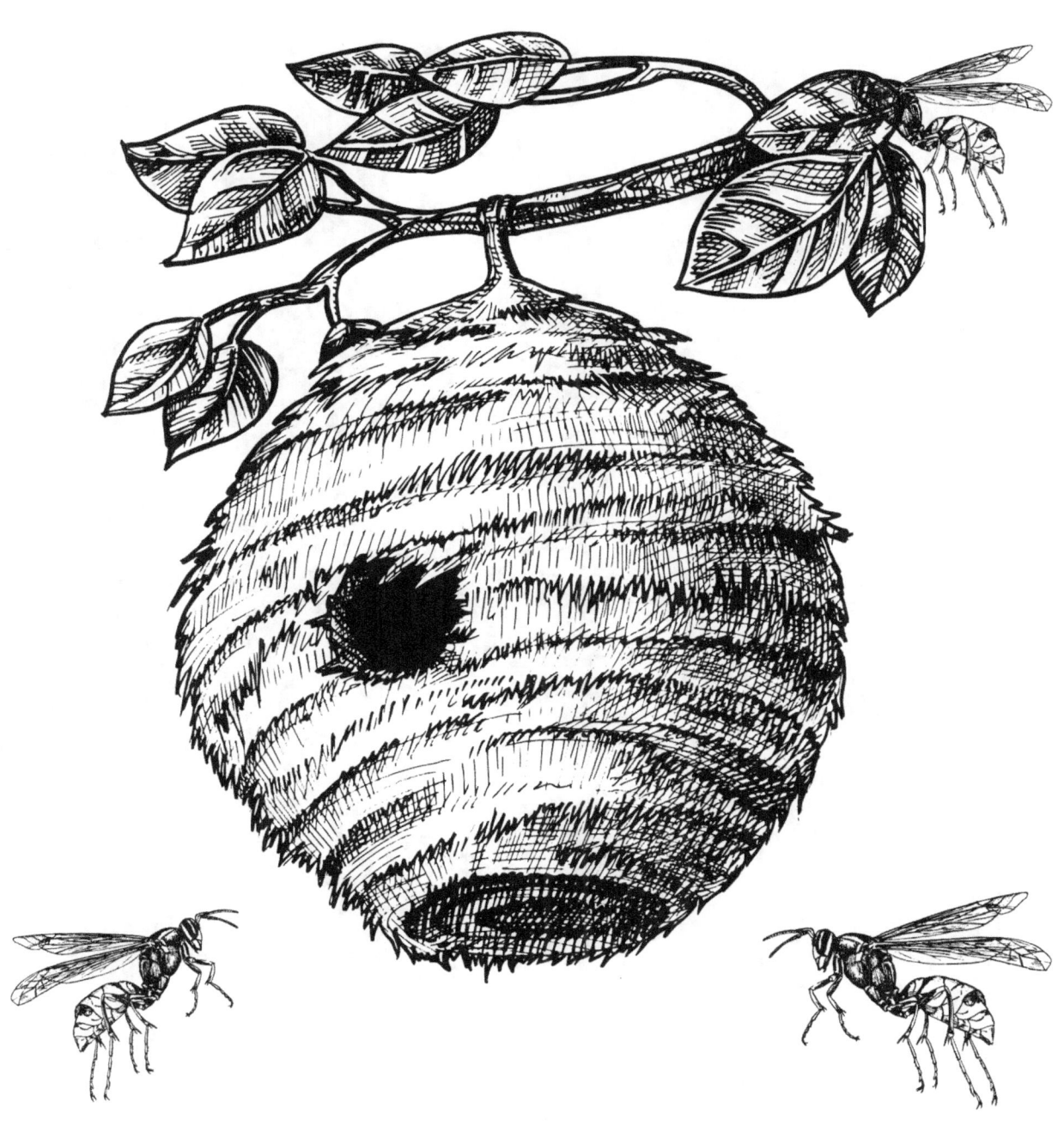

AGRICULTURE ECOSYSTEMS

A healthy agricultural ecosystem grows more food and keeps the soil and water clean.

WASP ECONOMICS

Wasp economics refers to the study of how wasps-especially social wasps - affect and interact with the economy and ecosystems in ways that can be analyzed using economic principles.

WASP ECONOMICS

While there's no formal field called "wasp economics," wasps play important roles that have economic value or cost.

WASP ECONOMICS

Wasps are natural predators of many insects that damage crops, such as caterpillars, aphids, grasshoppers, and flies.

WASP ECONOMICS

While bees are the main pollinators, some wasps (like fig wasps or paper wasps) help pollinate plants, contributing to crop production, biodiversity, and sustainable agriculture.

"Wasps may sting but they are a necessary part of our ecosystem."

-11 year old (Ardsley, New York)

NOTES